T0094316

# The Computer

# The Computer

## A Brief History of the Machine That Changed the World

Eric G. Swedin and David L. Ferro

An Imprint of ABC-CLIO, LLC
Santa Barbara, California • Denver, Colorado

**Library of Congress Cataloging-in-Publication Data**

Names: Swedin, Eric Gottfrid, author. | Ferro, David L., author.
Title: The computer : a brief history of the machine that changed the world / Eric G. Swedin and David L. Ferro.
Other titles: Computers
Description: Santa Barbara, California : Greenwood, an imprint of ABC-CLIO, LLC, [2022] | Revised edition of: Computers : the life story of a technology. | Includes bibliographical references and index.
Identifiers: LCCN 2022005417 (print) | LCCN 2022005418 (ebook) | ISBN 9781440866043 (hardcover) | ISBN 9781440866050 (ebook)
Subjects: LCSH: Computers—History. | BISAC: COMPUTERS / History | HISTORY / Modern / 20th Century / General
Classification: LCC QA76.17 .S94 2022 (print) | LCC QA76.17 (ebook) | DDC 004.09—dc23/eng/20220427
LC record available at https://lccn.loc.gov/2022005417
LC ebook record available at https://lccn.loc.gov/2022005418

ISBN: 978-1-4408-6604-3 (print)
978-1-4408-6605-0 (ebook)

26 25 24 23 22 1 2 3 4 5

This book is also available as an eBook.

Greenwood
An Imprint of ABC-CLIO, LLC

ABC-CLIO, LLC
147 Castilian Drive
Santa Barbara, California 93117
www.abc-clio.com

This book is printed on acid-free paper ∞

Manufactured in the United States of America

**Copyright Acknowledgments**

The authors and publisher gratefully acknowledge permission for use of the following material:

# Contents

| | | |
|---|---|---|
| Introduction | | vii |
| ONE | Before Computers | 1 |
| TWO | The First Electronic Computers | 27 |
| THREE | The Second Generation: From Vacuum Tubes to Transistors | 51 |
| FOUR | The Third Generation: From Integrated Circuits to Microprocessors | 69 |
| FIVE | Personal Computers: Bringing the Computer into the Home | 93 |
| SIX | Connections: Networking Computers Together | 121 |
| SEVEN | Moore's Law Triumphant | 149 |
| EIGHT | Social Media | 187 |
| NINE | Computers Everywhere | 219 |
| TEN | Information Security | 247 |
| Reflections on the Past and Future: A Conclusion | | 287 |
| Bibliography | | 289 |
| Index | | 295 |

# Introduction

The computer may be the greatest technological and scientific innovation of modern times. The computer has changed how we work, how we organize and store information, how we communicate with each other, and even the way that we think about the universe and the human mind. Computers have alleviated the drudgery of calculating sums and clerical work, and they have become essential tools in all organizations. Computers have become ubiquitous in many aspects of everyday life, and the future trend is that computers will become ever more powerful, more commonplace, and easier to use. This book tells the story of this evolution.

The story of the computer began long ago. Many ancient civilizations sought ways to automate mathematics. The clay tablets of Babylon, the Roman and Chinese abaci, the mechanical adders of Pascal in the 1640s, and the steam-powered devices imagined by Charles Babbage in the nineteenth century all led toward the modern computer.

World War II provided the impetus for the development of the electronic digital computer. After the war, the Cold War security and defense needs of the United States drove the development of computing technology. These advances principally occurred in the United States: the "giant brains" of the SAGE (Semi-Automatic Ground Environment) early warning system included an interactive interface, the connecting of computers across the country in ARPANET (Advanced Research Projects Agency Network) led to the internet, and the miniaturization of circuitry for use in space and missile technology encouraged the development of integrated circuits.

Billions of computers around the world serve us in many ways, from helping us to write books to microwaving our food. Only a couple of decades ago, most people could not imagine the value of a computer in their homes, while today a majority of households in Western society contain numerous computers in many different forms. Computers changed the workplace by making clerical work more efficient and raised the expectations for greater individual productivity. The role of the secretary has been reduced, as individuals within the workplace are now expected to master word processing, spreadsheet, database, and numerous other programs on their personal workstations. The use of these programs has reduced the expected turnaround time for any task that these programs facilitate.

The computer created a nexus through which two major trends in human development—advances in communication and automatic calculation—came together. With the development of digital circuitry, we see the digitization of the senses: motion, sound, the written word, and even tastes and smells, given the right technology. With the advent of the different digital networks that make up the internet, we see the possibility of vast volumes of digitized content moving across the globe in milliseconds. With networked devices, storing and using this information has become decentralized. The ease of manipulating digital content, either purposely or accidentally—especially text and pictures—unfortunately lends itself to fraud. As human production is digitized, identification and authentication practices struggle to catch up.

The computer has become such a powerful device that we often adopt it as a powerful metaphor. Much as the clock in the Middle Ages changed the way that people interacted with reality through measured time and Newtonian physics and the invention of the steam engine in the eighteenth century stimulated scientists to think of the laws of nature in terms of machines, the success of the computer in the later twentieth century prompted scientists to think of the basic laws of the universe as being similar to the operation of a computer. The new physics of information has come to view matter and natural laws as bits of information. So too did the computer change our way of thinking about *thinking*. Through their efforts to create artificial intelligence, scientists reimagined the mind in terms of computer resources and discovered new insights into the biological mechanisms of thought and memory, though actual thinking machines are still more fantasy than reality.

We have striven to write an accessible historical overview of this ever-changing technology, giving students and the curious lay reader an understanding of the scope of its history from ancient times to the present day. We illuminate the details of the technology while also linking those developments to the historical context of the times. This book is about the story of computers, but it is also the story of the people and events that drove the many technological innovations that led to modern electronic computers. Both authors have each spent over four decades in the computer field and have watched history unfold. We began our careers when punched card readers were still used and are still actively engaged in our exciting field.

The notion of *generations* in computing technologies seen in early chapters actually came from an IBM marketing campaign but was adopted in the historical field. The term *generations* implies an inevitable technological trajectory. It should be obvious after reading this book, however, that no technological trajectory is inevitable. With each innovation, the next innovation can appear more obvious, but innovations are also highly dependent on context. The case of artificial intelligence is most obvious in this respect, as you shall see.

In 2005, our book *Computers: The Life Story of a Technology* was published as part of the Greenwood Technographies series. Two years later, the Johns Hopkins University Press published a softcover edition. In the decade and half since, the history of computers has marched on, especially with the expansion of the internet; therefore, the story needed updating. This book has been revised and is almost twice the size of the 2005 book to accommodate an updated history of the computer.

We both want to thank our families for supporting us during this writing. For David: thanks to Marjukka, Stella, Robert, Jen, Luca, Andrea, Hugo, Miles, Michele, Lloyd, Michael, and Barbara. For Eric: thanks to my parents, Betty, Adam, William, Spencer, and Hannah. We also want to thank *Wired* magazine and the Wired website for such great journalism over the years, often telling the first iteration of stories as they became history.

# ONE

# Before Computers

## THE FIRST COMPUTER?

On Easter of 1900, a small group of Greek fishermen on their way home were pushed by a storm to Antikythera, a mostly uninhabited island north of Crete. Waiting out the storm, they did some sponge fishing in a cove on the island and found a large shipwreck. They notified the authorities, and archaeologists later discovered that the ship, possibly Roman, probably sank about 60 BCE. While the ship had long ago disintegrated, objects carried by the ship remained: many statues of bronze and marble, coins, and amphorae, all encrusted by coral.

Archaeologists also found a curious corroded lump of metal that seemed to be the remains of a mechanical device. Many thought it to be an astrolabe, a device useful in navigation. In the latter half of the twentieth century, with the help of x-ray photography and painstaking research, Yale professor Derek J. de Solla Price (1922–1983) discovered the machine's true purpose. It was a mechanical analog computer for calculating lunar, solar, and stellar calendars, and it probably contained more than forty gears. This was not a navigational device but likely a prized part of the cargo. This find changed the historical view of when such complicated and potentially powerful devices could have been created. Historians previously thought that the level of sophistication shown by the Antikythera mechanism was not reached until the medieval European astronomical clocks of the fifteenth century.

Astrolabes were more common than the unique Antikythera device. The ancient Babylonians originally divided the circle into 360 degrees

A part of the Antikythera device found off the coast of Antikythera near Crete. Scientists used x-rays to determine the device mechanically calculated celestial movements centuries before historians believed humans were capable of creating a device of this sophistication. (Marsyas)

and developed the twelve signs of the zodiac, which the ancient Greeks then used to create the astrolabe, probably between 200 and 100 BCE. An astrolabe is a circular device that maps the heavens. By rotating part of the device over the map, the user can potentially determine the current time, date, and latitude; determine positions of the sun and stars at any time of year; and calculate sunrise and sunset for any day of year. It can also be used to calculate heights, distances, and area related to the circle. It appears that the Antikythera device automated many of the relational calculations of an astrolabe through a series of gears and plates that showed the movement of the sun, stars, and moon.

At various times in the past, self-acting mechanical devices representing reality became very popular as both tools and objects of curiosity. Called *automata*, they probably existed before Philon of Byzantium (c. 280–220 BCE) created a washstand that automatically dispensed a pumice stone and a set amount of water for washing. In the first century of the Common Era, Heron of Alexandria (c. 75 CE) created automatic theaters with mechanical figures acting out the play *Nauplius*. In the seventeenth and eighteenth centuries in Europe and America, many mechanical devices allowed for serious measurement while using their gearing to enact little scenarios, or even whole plays, with mechanical figures of people, animals, and natural phenomena like thunder, lightning, or waves. One of the most popular devices related to the astrolabe was the orrery—a device that accurately showed the movement of the planets around the sun. One of America's earliest scientists, David Rittenhouse (1732–1796) of

Philadelphia, became famous in the eighteenth century for creating a beautiful and accurate orrery. Related to the orrery is the analog planetarium projection system that creates the image of the heavens on a domed ceiling; it is still used in planetariums the world over.

Another critical development in mechanical computing devices was the tide predictor. Entering or exiting a harbor can range from annoying to dangerous, depending on tides either running against your progress or taking your vessel dangerously close to submerged hazards. Creators of tables and charts did their best to prepare mariners with knowledge of tides by taking historical data and projecting them into the future. In the 1800s, Scottish physicist Lord Kelvin (1824–1907) created a formula for tides that he then modeled and refined with a machine that consisted of twelve pulleys (each pulley representing a coefficient of the equation) connected by a wire that was connected to a pen that drew the function (high and low tides) on a roll of paper. Each pulley was connected by a rod to a shaft that was turned to drive the machine. The gearing on the drive shaft could be changed to represent different locations on the earth. The U.S. Coast and Geodetic Survey created a similar machine with thirty-seven coefficients in 1911. That machine, 11 feet long and weighing nearly 2,500 pounds, was so successful that its accuracy (0.1 feet for any minute of a calculated year) was not matched until the mid-1960s by an IBM 7094 computer doing tens of millions of calculations for every year and location calculated.

The mechanical representation of mathematical formulas in the early twentieth century became most useful in what are called *differential analyzers*, which solved the problem of measuring the area under a curve. This problem can be so difficult to calculate that one technique used for a long period was to draw and cut out the curve on paper and then weigh the piece of paper—its weight was proportional to the area. A mechanical method developed by Lord Kelvin's brother James Thomson (1822–1892) was devised in the nineteenth century. His planimeter worked like a compass, with measurements created by the friction of wheels as they rolled along the curve. The necessity of moving the device carefully without slippage made it only marginally accurate, though it was not replaced until the introduction of more accurate machining technology in the 1930s. Vannevar E. Bush (1890–1974) at the Massachusetts Institute of Technology (MIT) created the first differential analyzer when he realized the speed it would give him in solving electrical power network problems, despite the long setup time of physically moving and rotating shafts for any particular differential equation. Bush directed the creation of three machines. The last was completed right before World War II, and it was used throughout the war to calculate ballistic tables. The complexity of the maintenance and programming of the Bush differential analyzer, and the machine's relative slowness in calculating ballistics, prompted the U.S. Army to invest in a machine that is considered the first modern computer: the Electronic Numerical Integrator and Computer, or ENIAC.

Bush's differential analyzer inspired a number of copies and became instrumental in World War II. A copy at Manchester University in Great Britain was constructed out of a children's erector set. Another, at the Moore School of Engineering at the University of Pennsylvania, was more sophisticated than the original. A number were created in Russia, Germany, and Norway. Early in the war, components of the Norwegian machines were either stolen or destroyed by Norwegian resistance fighters so that they would not fall into German hands.

The Antikythera device, astrolabes, and other measurement devices show the importance of calculation and modeling in human history. The human drive toward improving calculation and modeling, along with numerous mechanical and electronic inventions, combined to eventually create the modern computer. Much of these early efforts concentrated on creating multipurpose calculating machines to make mathematics easier. The story of mathematics is an important precursor to the rise of the modern electronic computer.

## MATHEMATICS

Along with language, mathematics has been a constant companion in human social evolution. Indeed, in many cultures, number systems were developed to a degree far greater than their use in basic needs. Often these numbering systems—such as those used by the ancient Mayans—were the province of the priestly class and used for complex religious ceremonies as well as having fun among themselves. The Greeks had two completely different systems: one used for numerical theory and another used for common purposes such as commerce. Esoteric mathematics has always been, and continues to be, known only by a few in society most highly trained in its use. However, over thousands of years, complex societies have required increasingly sophisticated mathematical skills at many levels of those societies. The trading of goods and services; the collection of monies for taxes; the building of structures such as pyramids, aqueducts, and skyscrapers;

measuring property boundaries; creating and implementing instruments of war; navigating across land and water; and understanding time continued to be critical areas that required the widespread use of mathematics.

The origins of numeration, or counting, in the human species are lost in prehistory. The best evidence we have for counting before the development of writing is linguistic. Various language remnants still exist showing the frequent use of the numeric bases associated with 5, 10, or 20—the fingers (and toes) on one, two, or all four human appendages. Tribes as geographically separated as the Inuit of Canada to various tribes in Indonesia have 20 as their numerical base. The Inuit used their word for *man* for each 20 units. Base 20 still exists in English in the word *score*, perhaps most famous in Lincoln's Gettysburg Address opening, "Four score and seven years ago."

Many cultures have not moved beyond number systems tied to specific objects. For example, in the 1960s, it was discovered that one language of the Indigenous peoples in British Columbia had different systems for counting people, animals, canoes, smooth objects, long objects, and round objects. Evidence of this still exists in English as well: a gaggle of geese, for example, where the word *gaggle* is only used to describe a large number of geese and nothing else. One of the indications of a more advanced concept in numbers is having an abstract number system that is not tied to particular objects.

At some point, in many locations around the world, humans moved beyond the use of fingers and pebbles for counting purposes and began recording counted values. Today, we have evidence of this through rock drawings or notches on sticks that often refer to groups of things (bison or other hunted game, for example). This developed into groups of numbers represented by symbols. The Romans used M = 1,000, D = 500, C = 100, L = 50, X = 10, V = 5, and I = 1. A later development allowed for the placement of the number to be used to attribute value. As early as 200 BCE, the Babylonians had a zero placeholder that they used to show a number as long as it had a nonzero digit in the ones (or units) place. Compared to our decimal number system of today, this is similar to being able to have a number like 107—where the locations of 1 in the hundreds place, 0 in the tens place, and 7 in the ones place make sense—but not being able to use zeros to pad out a number like 100. This inconsistency made the system less effective.

Our current decimal system, the Arabic system, began in medieval India. The earliest physical evidence of the Indian system is 595 CE, although it may have occurred much earlier. The Indian system allowed for more complex math. The movement of this new system appears to have occurred fairly quickly—most likely due to extensive trade. Persians borrowed the system, and evidence exists that they transferred the system to Europe through Spain in 976. The use of the new system in Europe, however, was rare until the 1200s. Leonardo of Pisa (1175–1250), better known as Fibonacci, advocated the Arabic system in a book titled *Liber Abaci* (A Treatise on the Abacus) in 1202. Translations and variations of the book *Arithmetic*, by the Arabic scholar al-Khowarizmi (780–850), finally succeeded in convincing great numbers of people of the usefulness of the system.

One reason the system was not widespread in Europe may have been the lack of Arabic translators for a number of centuries. The capture of Toledo from the Moors in 1085 may have helped solve the translator shortage. A number of merchants also resisted the new system. The city of Florence, Italy, prohibited the use of the new Arabic numbers in 1299 with the argument that they were too easily altered or forged. Roman numerals continued to be extensively used in Europe up through the seventeenth century, until the widespread adoption of the printing press proliferated and standardized the use of Arabic numerals and arithmetic. Roman numerals are still found in ceremonial forms today.

## EARLY AIDS TO MATHEMATICS

Some of the arithmetic technology that evolved along with human society began quite early in human history and has continued on to modern times. Given the cost of paper during much of human history, and the difficulty in working with clay or wax tablets, it is not surprising that other techniques were invented than just written ones—especially for determining intermediate results in calculations.

One technique that many societies used was tying various knots on cords to record numerical information. Biblical and Roman textual references indicate that those societies knew of the use of knots. Chinese records from as early as 2,800 BCE indicate that knots were used until at least 300 BCE. The Peruvian Incas in the sixteenth century used possibly the most sophisticated knotted string system ever known. This system, known as a quipu, consisted of a single string off which hung many knotted strings of different sizes and materials. The Peruvians recorded everything, from historical events to poems. German millers used a technique of knots to record flour sack contents until the beginning of the twentieth century.

Another technique relied on the use of tally sticks. Tally sticks are simply sticks or bones used as a surface for different carved markings. A tally stick known as the Ishango bone, 8,000 years in age, has been found in Zaire and represents a six-month lunar calendar. Other bones with markings have been found that are up to 37,000 years old, but they are less obviously tally sticks. The Chinese used tally sticks, and the remnants of those times remain in their language—the Chinese character for *contract* is "large tally stick." The British government used tally sticks from the thirteenth to the nineteenth centuries. All contracts were recorded by the Exchequer, and the Exchequer tally system included cutting various notches into a stick and then splitting the stick from end to end. The bank kept one side, called the *foil*, and the individual kept the other side, called the *stock*. This is where the term for owning part of a company as *stock* originated as well as the term *tally up*. The centuries-old collection of sticks was finally deemed obsolete in 1826. In 1834, the British government began the process of burning the sticks in the stove of the House of Lords. Unfortunately, the sticks burned too hot and lit the paneling on fire, and soon the House of Lords and the House of Commons had burned to the ground.

Knotted strings and tally sticks were probably used more for storing information than calculation, though they could be used in conjunction with other techniques to calculate. For example, the Chinese were probably the first to have a complete decimal system of numbers dating from as early as 1300 BCE. One technique the Chinese developed soon after the development of the decimal system was calculating rods. Rods could be made of wood, bamboo, bone, or ivory. Ivory rods were the most expensive and exclusive. These rods were combined in different ways to form the necessary numbers. The origins of the abacus include the use of these calculating rods. Later on, the rods were laid out on a board or cloth that was divided into squares. Each square would be a different digit of a number. By 800 CE, they had added the zero to this system. The Chinese also had black rods to represent negative numbers and red rods to represent positive numbers. Red is a lucky color in Chinese; the term *in the red* would have the opposite meaning to what it means in the West. The Koreans and Japanese adopted the rods as well. The Japanese solved the problem of the rods rolling into the wrong square by flattening the rods.

The rods were displaced by the modern form of the wire and bead abacus sometime around 1300 CE in China. The abacus seems to have had its original origins in the Middle East, possibly modern-day Turkey or Armenia, during the Middle Ages. Tabular or board abaci existed in other regions much earlier. Because surviving Greek writings are more about their theoretical theorems than common mathematics, we do not have much evidence about how the ancient Greeks did calculations. However, the Salamis abacus found on the Greek island of Salamis, an approximately two-by-five-foot marble table with numerical demarcations, is an excellent example of an early abacus. References in Greek and Roman literature and paintings also place the first abacus as early as 400 BCE. At first, stones were used as counters on these boards. Later, in medieval Europe, coin-like counters were used.

The French introduced *jetons* in the fifteenth century, commemorative sets of counting coins that were often given as gifts on New Year's Day. Jetons were exported to the French colonies in North America until 1759, when the French suffered defeat at the hands of the British during what is known as the French and Indian War. Interestingly, the French took up the use of paper and pen using Hindu-Arabic numbers and abandoned the use of jetons soon after and by the 1812 invasion of Russia by Napoleon, French soldiers were bringing back commandeered Russian abaci and jetons as curiosities.

According to the Jesuit priest Joseph de Acosta (1540–1600), in Peru in 1590, some form of abacus existed in the Americas as well. Unfortunately, little is known about the Peruvian device. In the right hands, the abacus could be manipulated quite quickly. In 1946, during the American military occupation of Japan, two individuals, Kiyoshi Matsuzaki of Japan, using the Japanese soroban abacus, and Thomas Wood of the U.S. Army, using an electromechanical calculator, squared off in a contest of calculation speed and accuracy. Matsuzaki won.

Another technique invented in many locations has been called *finger calculation*. Evidence suggests that finger techniques existed prior to 500 BCE in Greece. Because of the needs of trade across many cultures, an informal standardization of finger calculation likely occurred early in human cultural interaction. This allowed bargaining without the need to learn another language. A complete description of the system of finger calculation that was likely used from Europe to China was written by the Venerable Bede (673–735), an English monk, in 725 CE. The system used the left hand to represent from 1 to 99 and the right hand to represent from 100 to 9,900. Apparently, there were signs using the rest of the body that could represent up to 1,000,000, but they are not detailed as well. The system allowed for holding intermediate values on the hand as a temporary "register" while calculating in the head. Other finger systems existed as well. One system used in Europe through the 1700s could accommodate the well-educated people who still seldom knew the multiplication tables beyond 5 times 5 and allowed multiplication of the digits from 6 to 10. This system was taught in some Russian schools into the 1940s.

The Greek astrolabe was likely originally designed for calculating time and location. However, it was also a device that could be used for calculation that used the circle as a reference. Three later classes of devices also utilized these principles: the quadrant, the compass, and the sector. All three devices were used in one form or another into the twentieth century.

The Gunter's quadrant is an astrolabe consolidated onto one-quarter of a circle by Edmund Gunter (1581–1626) of Gresham College in London in the seventeenth century. Many quadrants were created with a number of scales on them, including trigonometric functions as lengths of lines. These were used in conjunction with trigonometric tables. Many also had squares and cubes and their roots on their backside. Large books were written describing all the many functions that a quadrant could do.

The proportional compass was a set of dividers with the hinge between the two legs being an adjustable and scaled point—usually between one and ten. The scales allowed for obtaining squares or square roots geometrically using lines, circles, or solids. For example, set the scale at four and measure a square of one-inch a side on the small end of the compass and you automatically obtain a square that will have four times the area on the other end. The origination of this device is lost, but documentation existed in the sixteenth century.

A sector was two scales hinged at one end. The Italian scientist Galileo Galilei (1564–1642) created one of the first around 1597. The initial use of the sector was by the military to calculate gun trajectories—not the last time an advanced calculation device was created for such a need. The device could generally measure an angle between zero and ninety degrees for a gun's elevation. Zero degrees was left blank on the device, and so the term *point blank* came into existence. Later versions of the sector combined it with the compass and also had a curved interior to allow for measuring the size of the cannon ball. The number of scales also included cannon diameter or caliber, shot weight, amount of charge needed, and more. These devices also allowed for more generic calculations and could be used in conjunction with other sectors and various tables to do quite complex equations. The devices required a fair amount of skill in handling for advanced functions, leaving ample room for further improvement. In addition, precalculated tables were notoriously error prone, which also drove the need for more robust calculating tools. Tables were necessary, however, to simplify complex mathematics. Through a number of techniques, tables often turned difficult-to-calculate equations, including multiplication and division, into simpler-to-use equations of addition and subtraction.

## JOHN NAPIER'S BONES

John Napier (1550–1617), born in Merchiston Castle, Scotland, is credited with creating at least two critical developments in the evolution of computation: logarithms and "Napier's bones." Born at the beginning of the Scottish Reformation, Napier spent much of his time managing the family's estates and engaging in radical theological musings. He also found time for mathematics, and while some historians have speculated that logarithms were independently invented elsewhere, there is little evidence of such invention, or at least little evidence that it was communicated to Scotland.

Napier discovered that a series of numbers could be found that had a corresponding series where the numbers were what he termed *logarithms.* His 1614 book, *Description of the Admirable Cannon of Logarithms*, known in its original Latin as the *Descriptio*, described logarithms and included a series of tables with the logarithms of many numbers. This concept quickly spread, and a number of more complete tables by other mathematicians followed. Henry Briggs (1561–1630) published a book of tables in 1624 that included the logs of numbers from 1 to 20,000 and from 90,000 to 100,000 to 14 decimal places. Johannes Kepler (1571–1630), Edmund Wingate (1593–1656), Adriaan Vlacq (1600–1667), and Edmund Gunter (1581–1626) all added to and recalculated tables of logarithms.

To create the *Descriptio*, Napier invented another device to aid in calculation that he called the *Rabdologia*, which consisted of a set of plates that could be organized with respect to one another to give a multiplication product. The idea came from a more ancient method of multiplication called *gelosia*, where a matrix of multiplicands was created. Napier's Rabdologia became best known as "bones" because the best sets were created from ivory. Napier's bones only became known after his death in 1617 because Napier did not believe them worthy of publication. Once published in *Rabdologia*, however, many inventors furthered his work by creating more sophisticated sets of "bones." Gaspar Schott (1608–1666), a Jesuit priest in Rome, published the ideas of Athanasius Kircher (1601–1680) that extended Napier's bones with a creation he called the *Organum Mathematicum* in the 1660s. The Organum included bones for things as diverse as addition, subtraction, geometry, calendars, spheres, planetary movement, the construction of canals, the construction of military fortifications, and music. Schott also invented a box that put Napier's bones on cylinders, which ensured that the bones were correctly lined up and made calculation easier. The Englishman Samuel Morland (1625–1695) independently invented the same type of device in the mid-1660s. The technique of lining up Napier's bones in a mechanical way became a milestone in the creation of mechanical calculating machines. The most advanced version of Napier's bones was the Genaille-Lucas rulers created by Henri Genaille (d. 1903) and Édouard Lucas (1842–1891) in 1885. These rulers eliminated the problem of the carry value between partial products.

Napier's bones also became instrumental in a device so potentially efficient and inexpensively produced that it later served engineers as their primary calculating device until the early 1970s: the slide rule. In 1620, Edmund Gunter created a scale of logarithms on a stick of wood that could be used with a pair of dividers to easily add logarithms. The Englishman William Oughtred (1574–1660) simplified this invention by eliminating the need for dividers by having two scaled pieces of wood slide past each other, and thus he became the father of the modern slide rule in 1622. Students of Oughtred created variations of the slide rule—one publishing before Oughtred did because Oughtred did not believe the work worthy of publication. A number of improvements on the slide rule continued over the next two centuries, but the device did not really replace the use of dividers and scales until the Scottish inventor James Watt (1736–1819) created an inexpensive and accurate slide rule in the late 1700s and the nineteen-year-old Frenchman Amédée Mannheim (1831–1906) added the idea of the moving cursor over the slide in 1850. Mannheim's eventual appointment to professor at the École Polytechnique in Paris helped the slide rule become a mainstay device for mathematics.

## MECHANICAL DEVICES

The eventual creation of the electronic digital computer drew heavily on the first mechanical automation of calculations. Mechanical calculators needed multiple parts: mechanisms to enter the number into the machine and select the correct motions for the correct function; a means to store the temporary values within the machine, including a possible carry value; a method to display a result; and a method to reset the machine to zero. This all needed to occur with a minimum of human intervention. Unfortunately, early efforts at mechanical calculators suffered from less developed skills in machining and manufacturing. New gearing and techniques also needed to be invented to satisfy the tolerance requirements of the machines. The most skilled machinists, often found in the watch trade, were also not necessarily available for doing the work due to an initial lack of paying customers.

Although often attributed to Blaise Pascal (1623–1662), the first mechanical calculating machine probably belonged to Wilhelm Schickard (1592–1635), a professor of math, astronomy, geography, Hebrew, and Oriental languages (as well as a Protestant minister), in Tübingen, Germany. Schickard became fascinated with Kepler's descriptions of Napier's bones and created two machines that automated the multiplication process. While both machines are now lost, in 1971, it became possible to reconstruct the machine from some of Schickard's notes. The machine worked, but a problem with early carry mechanisms became clear from the reconstruction. The gearing would potentially damage the machine if a carry needed to be propagated through the digits, for example, adding 1 to 9,999.

The first calculating machine for which a copy still exists today came from Blaise Pascal. He worked out Euclid's geometric theorems on his own at age twelve, described complex conical geometry in a treatise when he was sixteen, and worked with Pierre de Fermat (1601–1665) to establish probability theory. In 1642, at the age of nineteen, Pascal invented a mechanical device for adding and subtracting, which he called the *Pascaline*, to assist his father in his job as tax collector of Normandy, France. To create the machine, Pascal had to train himself as a mechanic because local mechanics did not work to the fine precision needed for this machine. The machine used gears and wheels similar to a mechanical odometer in modern automobiles. The numbers appeared in small windows, and below those windows were dials similar to those on a rotary telephone. The operator used a stylus to turn the dials. The number windows actually displayed a choice of two values. If you examined the top number, you would have addition. Subtraction was done by observing the bottom number, which was the nine's complement. Pascal's mechanism did not allow for subtraction, as the gears could not run in reverse. However, the gearing did

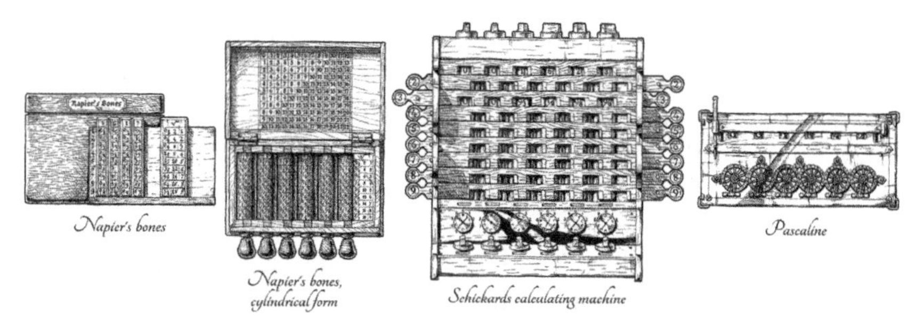


Illustration of Napier's bones in two forms, Schickard's calculator and the Pascaline. (Alexander Babich/Dreamstime.com)

eliminate the carry problem that was obvious in Schickard's machine. Over the course of his life, Pascal made almost fifty different versions of the machine. All of them were fairly temperamental and required constant maintenance.

In 1694, Gottfried Wilhelm Leibniz (1646–1716), a German rival mathematician of England's Isaac Newton (1642–1727), created a machine called the *Leibniz wheel* that worked similarly to Pascal's and also did multiplication. His first demonstration was in 1672, when he showed a wooden version that did not work very well to the Royal Society in London. A metal version was created in 1674 with the help of clockmaker M. Olivier. The machine introduced the concept of the stepped drum, a gear that has progressively deeper teeth, as a means to select the correct number. Unlike Pascal's machine, Leibniz's machine could move in reverse. Unfortunately, propagating a carry often required the user's intervention, so the machine signaled the user when a carry was needed.

Samuel Morland realized his own penchant for mechanical work when visiting the court of Queen Christina of Sweden in 1653, where he saw one of Pascal's adding machines. Morland worked on a version of his own and published his designs for three different machines in 1673. One machine was a mechanical adder. He elected not to create the complicated carry mechanism of Pascal or use the technique of Schickard. His machines were similar to Leibniz's in that they indicated to the operator when a carry propagation needed to occur. Because of this, they were simple to operate and reliable and could be made quite small; extant examples are three by four inches and only a quarter of an inch thick. Despite this convenience, few were sold. Another of Morland's machines automated Napier's bones.

It allowed for the replacement of disks—a set of thirty in all that were essentially circular versions of the bones—and could find squares and cubes and their roots as well as perform multiplication and division.

The invention of a commercially successful automatic adding machine had to wait until the nineteenth century. In 1820, Charles Xavier Thomas de Colmar (1785–1870), serving in the French army, invented what he called the *arithmometer.* The arithmometer was a mechanically improved version of Leibniz's wheel. The size of a tabletop, the device could add, subtract, multiply, and divide. After leaving the army, de Colmar joined the new field of insurance. Using the calculation of insurance tables as an incentive, he continually improved the device, submitting it to various scientific competitions, and won the Legion of Honor. Some later devices had as much as a thirty-two-digit product register. Variations of this machine were used through World War I.

The simultaneous invention of a variable-toothed gear to replace the Leibniz stepped drum by both Frank S. Baldwin (1838–1925) in the United States and Willgodt T. Odhner (1845–1905) in Russia resulted in a considerable reduction in size and weight for calculating machines. Now calculating machines could sit on the corner of a desk. The Brunsviga company in the United States began manufacturing these machines in 1885 and sold 20,000 of them in the next three decades. Other equally successful companies also manufactured these types of machines. Mechanical improvements helped drive the market, including more functions and reduced size, and the comptometer, invented by Dorr E. Felt (1862–1930), improved the speed at which the keys could be pressed without the machine jamming. Until the advent of the digital calculator, these types of machines were the backbone of automated calculation on desktops in the Western world.

## CHARLES BABBAGE

Charles Babbage (1791–1871) was born in England. He showed an interest in both the internal workings of mechanical things and advanced mathematics as a child and had become a well-known mathematician by the 1820s. While a student at Cambridge University, Babbage and two friends translated, annotated, and added interesting examples to a French text on calculus that became the standard text for calculus instruction in Britain for most of the nineteenth century. After leaving Cambridge, Babbage spent some time traveling, including a solo descent into the crater of Mount

Vesuvius to spend the day taking numerous air pressure and temperature measurements while dodging venting hot air and lava. After moving back to London, he invented a cowcatcher for the steam engines of the British railway system and an air-conditioning system for his own London apartment. Babbage inherited a moderate fortune and did not have to worry about income for most of his life. A charming and busy member of London society as a younger man, he became ever more reclusive with age as he obsessed over his two greatest inventions: the Difference Engine and the Analytical Engine. Babbage also became a permanent enemy of London street musicians when he argued to have them outlawed. For the rest of his life, he was hounded by street musicians, who apparently took many opportunities to gather under his window in the middle of the night and play for him.

In 1826, Babbage cemented his reputation in mathematics by producing a table of logarithms that were the most complete and accurate yet published. Babbage not only focused on the correctness of the tables but also on their readability. He experimented with different type settings and the colors of type and paper. Some tables of logarithms at the time had more than 1,000 errors. The errata pages published to correct them often introduced even more errors. Despite his best efforts, Babbage's tables still ended up with approximately forty errors between his manuscript and the final printing. This did not satisfy Babbage, who felt that a machine might be the only way of removing the inevitable human errors.

After discussions with his friend, astronomer John Herschel (1792–1871), Babbage imagined reducing the errors present in some mathematical tables used by astronomers by creating a machine—possibly powered by steam—to do the calculations. He imagined a difference engine. The method of differences takes into account that values placed into the variable of a polynomial will create a constant difference in the value of the function. With simple functions that have a single term, the technique might require a single iteration through the equation. More complicated functions would require more iterations where the differences of the differences would have to be found. A difference engine that would automatically cycle through the differences had already been imagined by Johann Helfrich Müller (1746–1830), an engineer in the Hessian army who had created a mechanical calculating machine. Müller's ideas were included in a book written by E. Lipstein in 1786, but he had failed to find funding for further development.

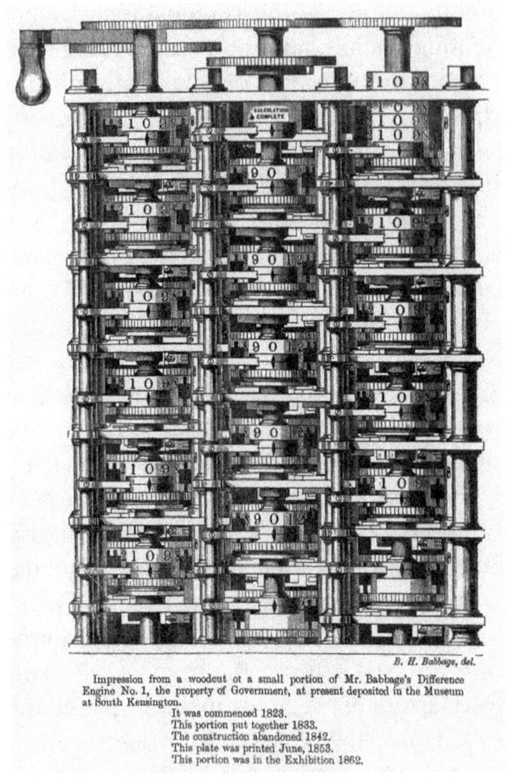


Woodcut print of a part of the Babbage's Difference Engine. (Public domain. From Charles Babbage, *Passages from the Life of a Philosopher.* London: Longman, Green. Longman, Roberts & Green, 1864)

In the 1820s, Babbage and many of his friends began to communicate with the Royal Society and the British government in support of his project to create a difference engine. As a demonstration, Babbage built a six-digit working model of the Difference Engine that could calculate to two levels of differences. At that time, the British government had a 200-year history of funding practical science and technology, and the government was interested in the Difference Engine to assist in creating actuarial, tide, navigation, engineering, logarithmic, and interest tables—all useful for the British Empire. The government advanced Babbage £1,500 in 1823, and Babbage agreed to personally fund the remainder of the project up to £5,000.

Babbage soon realized that the kind of machining necessary for the project would require expertise, and he found such an expert in Joseph Clement (1779–1844). In pursuing the complex mechanisms needed for the engine, Babbage and Clement advanced the level of machine knowledge in both mechanisms and tooling in Britain. Many of the machinists who apprenticed in Clement's shops became quite successful. One, Joseph Whitworth (1803–1887), created the Whitworth standard for nuts and bolts. This is an early example of government investment in a technological project paying itself off in unexpected dividends. Unfortunately, the primary goal of creating the Difference Engine became complicated. Babbage lost his wife, daughter, and father, and his own health deteriorated.

He moved to Italy for a while to recover. Delays in getting funding from the government also created work stoppages. Babbage was forced to let staff go, and then, hire and retrain them for each stoppage. He finally built a shop on his own property to house Clement and his staff, but Clement became concerned over the working arrangement and conditions. Their relationship deteriorated. Clement decided to leave, and British law supported him in taking all the tools with him.

During one of the stoppages, Babbage redesigned the Difference Engine, and he applied for further funding to create the new version. The British government did not see the redesign in a positive light and canceled any future funding in 1842. Babbage handed the plans and completed components of the first Difference Engine to the government. When completed, the machine would have had eighteen-digit numbers and have calculated to six differences. The machine would have automatically corrected round-off errors. It also had built-in safety features that would have stopped the machine from damage if it had any malfunctions. It would have printed the tables directly as well so as to avoid any errors in print setting. In 1991, the Science Museum in London took Babbage's detailed plans for his redesigned Difference Engine and built it: it was ten feet long and six feet high, weighed three tons, and contained 4,000 parts. The Difference Engine worked with only minor changes in the design. The components for the first Difference Engine and a recreated second Difference Engine sit in the Science Museum today.

A number of difference engines were attempted by others in the following hundred years. Inspired by Babbage, the Swede Georg Scheutz (1785–1873) and his son Edvard created a small machine that could find three differences in 1843. In 1851, the Swedish government funded a complete machine with the stipulation that it be completed before the end of 1853. Their machine was completed by October of that year, and the Scheutzes improved its capability with an additional grant in 1854. In that year, the machine was examined by the Royal Society and praised by its members, including Babbage. After the machine won an award in Paris, it was purchased by the Dudley Observatory in Albany, New York, and was later purchased by Dorr E. Felt, inventor of the comptometer adding machine. Today, the Scheutz difference machine is located at the Smithsonian Institution in Washington, DC. A second duplicate machine was created by the Scheutzes in 1856. It was funded, ironically, by the British for £1,200 to assist the Astronomer Royal and the insurance industry. Today, that machine is located in the Science Museum in London. Another Swede, Martin Wiberg (1826–1905), created a smaller and lighter version of the Scheutz machine in 1860. Others were also built, but by the 1930s, inventors realized that

Recreation of Charles Babbage's Difference Engine Number 2 (or the Analytical Engine) at the Science Museum, London. Created in 1991, the bicentennial of Babbage's birth. It is considered the first attempt at a general-purpose computer. (Massimo Parisi/Dreamstime.com)

common desktop mechanical calculating machines could also be the basis for a difference engine.

While building the Difference Engine, Babbage began to imagine a different machine that he called the Analytical Engine. This new machine established the major logical components and techniques for the modern electronic computer, though it was never built. Today, the proposed Analytical Engine is considered the first realizable design for a general-purpose computer. Its body was composed of three main components similar to today's computer's central processing unit (called the "control barrel" by Babbage), arithmetic processing unit (called the "mill" by Babbage), and memory (called the "store" by Babbage). The machine also had an input device for loading a program, using Jacquard punched cards, and an output device that printed results. Programs could have controlled repetitive operations (called *program iteration* or *program loops*). An 1840 version of the design would have stood fifteen feet high. The circular mill and control barrel structure would have had a radius of six feet, and the store would have extended out from this structure ten to twenty feet. Card readers would have added to the size.

The control barrel (actually three different barrels) was similar to a music box cylinder. The studs on a music box pluck the right musical note. The studs on the control barrel pushed various control rods. The rods connected to other parts of the machine, so the barrel acted like the microcode of the machine. It created the processing sequence, directing the machine as to when to read from and write to store (memory), when to read from the program, and when to move register information to the mill, along with any other necessary operations. Information from and to the registers passed along a rack of gears that simply transferred the settings of the individual registers to the mill. The mill then performed the correct mathematical function, as directed by the program, on the numbers given to it.

The store had become obvious to Babbage as he constructed the Difference Engine. He realized that the registers that held each number to be manipulated were essentially identical. Instead of limiting the registers for particular functions, he created the store so that the registers could be used as a program directed—essentially creating the concept of random-access memory (RAM). In addition, the memory was extendable. It was designed so that more registers could be added. The most detailed design by Babbage called for sixteen registers that could store either one twenty-digit number or two ten-digit numbers. Other plans showed that Babbage may have envisioned fifty register columns of forty-digit numbers. There was also a special register that allowed for counting how many times a programming operation was performed—useful for doing program iteration.

Programs were input into the machine by the use of Jacquard punched cards. Jacquard cards were rectangular, made of hard paper stock, and linked together with cloth along their long edges. The Frenchman Joseph Marie Jacquard (1752–1834) essentially invented the first looping programmable machine in 1801. The machine was not used for calculation, however, but as a loom for weaving tapestries. Punched cards for weaving had previously been invented by Jacques de Vaucanson (1709–1782) in 1745, but his cards needed to be fed into the machine one at a time. What was later called the *Jacquard loom* used connected cards on a belt on rolling drums. The cards were automatically fed into the machine, and as the cards passed a certain part of the machine, rods descended. The rods that could pass through the holes in the cards selected a thread to be woven into the tapestry. This meant that a deck of cards could be used repeatedly, creating the same pattern. Decks of over 24,000 cards that could create elaborate patterns were eventually made. Napoleon awarded Jacquard for his work, and his business flourished. The professional weavers of the day, however, did not see the machine in quite the same glowing terms: Jacquard's life was threatened, and some machines were destroyed by vandals. Some versions of the machines remain in use today.

Babbage envisioned a number of Jacquard card readers attached to the Analytical Engine, with some readers potentially able to control the flow of others. The mechanism allowed reading the cards in sequence as well as out of sequence with the reading mechanism able to reverse the flow of cards as well. This feature meant that certain parts of the program might be repeated or skipped—thus establishing the three main types of advanced programming language statements: sequential, iterative, and conditional. A simple program on five cards might work as follows: store 2 in V1, store 3 in V2, read V1, read V2, multiply. The answer would print out as 6.

Babbage chose to think of the Analytical Engine as an academic exercise. Not one of the components designed was ever built by him, though the detailed parts of the final design seem to have been scrupulous in accommodating the machining techniques of his day. The parts would have required tolerances of one by five hundredth of an inch—possible, though expensive, in the mid-nineteenth century. Babbage continued to add to and modify the design for the remainder of his life, although only modifications to the main design occurred after 1840.

Some important details of the machine were written by other people excited by the project. While in Italy on a trip to Turin, Babbage described the machine to a number of interested engineers. Luigi F. Menabrea (1809–1896), a military engineer and future prime minister of Italy, wrote about the machine's operations in Italian. A friend of Babbage, the daughter of the poet Lord Byron, Lady Ada Augusta Lovelace (1815–1852), translated Menabrea's text into English and added considerably to the machine's operational instructions. Ada Lovelace has been called the "first programmer" for her efforts. In 1906, Babbage's son, Major-General Henry P. Babbage (1824–1918), had the R. W. Monro company create a version of the mill part of the machine to prove that it would work. It did, printing (using twenty-nine decimal places) the first twenty-five multiples of pi.

## THE HUMAN COMPUTER

In the 1820s, while Babbage worked to create the Difference Engine, he traveled through Europe looking for manufacturing techniques that would suit his machine. He did not find much to add to the machine. However, in the process, he became an expert in the latest manufacturing techniques in Europe. As a trained economist, he saw these techniques in not only engineering terms but also economic terms. In 1832, he wrote the classic economic text *Economy of Manufactures*, which along with the 1776 text *Wealth of Nations* by Adam Smith (1723–1790) were important sources of intellectual information as industrialists tried to understand the social forces that created the Industrial Revolution. The scientific management techniques of Frederick W. Taylor (1856–1915) in the late nineteenth and early twentieth centuries also relied on the insights of Babbage and Smith. Ironically, though Babbage wanted to remove the human element as much as possible from creating mathematical tables, his economic ideas were used to more efficiently industrialize information processing through the use of even greater numbers of people doing calculations. In the nineteenth century and up through World War II, the term *computer* referred not to a machine but to a person. Human computers used devices such as the slide rule, abacus, and pen and paper; later, they used electromechanical devices like adding machines.

The first truly organizational approach on a massive scale to the creation of mathematical tables was probably that of Baron Gaspard de Prony (1755–1839) in 1790. In an effort to reform the tax and measurement systems after the French Revolution, Napoleon charged de Prony with creating these new tables, a huge undertaking. Inspired by Adam Smith, de Prony organized the effort along the lines of a factory with three sections. One section employed prominent mathematicians who determined the mathematical formulas to be used for calculating the tables. Another section used these formulas to organize the people who would do the calculations and collated the results. The third group consisted of up to eighty people who actually did the calculations. The calculations were made using the difference method, the same method that the Difference Engine used to break complex equations into equations with only addition and subtraction. This simplification allowed de Prony to hire less educated individuals for the third group. In fact, de Prony hired, in his own words, "one of the most hated symbols of the ancient regime": the former hairdressers of the elaborate powdered wigs of the aristocracy.

Nothing on the scale of the French effort was known to exist in Babbage's England until the time that he wrote *Economy of Manufactures*, in which he included a description of the Bankers' Clearing House of London. The increasing popularity of bank checks pushed the Bankers' Clearing House—a secretive organization that Babbage managed to gain entry to through subterfuge—to prominence in London's financial circles. Almost

£1 billion in 1830s currency was exchanged there every day. The sophisticated clerical organization allowed these exchanges between bankers. After the 1850s, this financial infrastructure model became more prevalent in Great Britain, and by 1870, the Railway Clearing House had over 1,300 clerks. In 1875, the Central Telegraph Office had over 1,200 clerks and four times that many in 1900. The rise of savings banks as a place to deposit discretionary income from the burgeoning industrial working class also increased the demand for clerical organizations skilled in processing a large number of transactions with minimal errors. For the first time, companies could offer insurance policies to working-class individuals because their processing organization made the administrative cost of individual policies less expensive.

## HERMAN HOLLERITH

No commercial entity in the United States approached the proportions of the European companies in terms of clerical needs in the late nineteenth century. However, one governmental organization did, the U.S. Census Bureau. In 1790, an act of Congress determined that a census of the population should occur every ten years so as to apportion members of the House of Representatives, one representative per 33,000 people. While the bureau employed fewer than thirty clerks in 1840 when the population was 17 million people, the bureau, along with the population, grew quickly after that. By 1880, the bureau employed almost 1,500 clerks to count a population close to three times that of forty years before. The census consisted of an army of census takers that created a huge amount of paper forms; these included information about individuals, such as their sex, age, ethnic category, location, and so on. Back in Washington, DC, the forms were collated on large tally sheets, a grid with rows and columns that corresponded to the various categories. Clerks created the final statistics after many passes through the forms. The whole process was done by hand with pen and paper.

In the 1880 census, the director of statistics, John S. Billings (1838–1913), suggested to a young staff member, Herman Hollerith (1860–1929), that the counting of the census should be automated by mechanical means. Hollerith had been a student at Columbia University's School of Mines. Between the 1880 and 1890 censuses, Hollerith worked at the U.S. Patent Office and as an instructor in mechanical engineering at the Massachusetts Institute of Technology (MIT). In his spare time, he constructed a prototype mechanical tabulating machine in anticipation of the 1890 census. By 1888, census officials concluded that another technique for the census had to be arrived at. The bureau had taken seven years to tally up the 1880 census, and it was feared that the 1890 census might not be completely tallied until after the 1900 census.

The new superintendent of the bureau, Robert P. Porter (1852–1917), sponsored a contest to replace the old system of pen and paper. Three individuals, including Hollerith, vied for the census contract. Two of the systems proposed were still human centered. They used different colored ink or cards and other techniques to make the process easier. Hollerith created a mechanical system inspired by his travels on railroads. When a train conductor took a ticket from a patron, he punched holes in the tickets that corresponded to where the patron was sitting and the physical characteristics of the patron, like hair color and gender. It occurred to Hollerith that this system would work for the census.

Hollerith proposed that clerks take the reports from census workers and create a paper card representing each individual in the United States by punching out the correct characteristics. The cards would then be tabulated automatically by a machine that would tally the characteristics by census district. The machine worked by passing metal rods through the card holes, a method similar to the Jacquard loom. In Hollerith's case, however, the rods dipped into small cups of mercury. The rods that made contact with the mercury passed an electric current through and thus incremented electromechanical counters mounted on the front of the machine. The bureau contest was held in 1889 and consisted of reprocessing the St. Louis returns from the previous census. The three systems were fairly comparable in terms of speed when it came to taking the census reports and putting them on cards. After that phase, however, Hollerith's system was obviously so much faster and more flexible that he won the contract for the 1890 census.

Workers at the bureau created over sixty million cards for the census. While some of the newspapers were skeptical of the Hollerith machines, the bureau knew it had a winner. The bureau took only two and a half years for the census to be counted at a cost of $11.5 million, $5 million less than the expected cost if the work had been done by hand. Hollerith's machines were improved and used again for the 1900 census. In 1896, Hollerith created the Tabulating Machine Company (TMC) and sold or leased the machines to other countries for their census operations. The machines also began to find their way into the private sector and were used to compile statistics for railroad freight, agriculture, and more. In

Herman Hollerith's Tabulator for the U.S. Census in 1890. The tabulator used punch cards to store data, and they could be tabulated one at a time by using the shown mechanism, which created an electrical circuit that incremented the shown radial counters. (Photobulb/Dreamstime.com)

1911, TMC merged with another company and became the Calculating Tabulating Recording Company (CTR). In 1910, an aggressive and successful salesman named Thomas J. Watson (1874–1956), who worked for the National Cash Register Company (NCR), saw the potential of the CTR technology. By 1915, Watson led CTR. He focused the company on big leasing contracts, staying away from small office equipment sales. In 1924, he renamed the company International Business Machines (IBM).

IBM was one of the four business machine companies that dominated office equipment sales in the first half of the twentieth century in the United States. Of the four—IBM, NCR, Remington Rand, and Burroughs

Adding Machines—IBM was the smallest. By 1968, though, IBM was bigger than the other three combined. In the late 1930s, IBM invested in the electromechanical Harvard Mark I, a realization of Babbage's dream of an Analytical Engine. Most of the calculations done in the twentieth century up through World War II, however, were not done by sophisticated general-purpose machines, for they did not exist yet. Equipment such as IBM's tabulators and Burroughs' electromechanical adding machines fulfilled the increasing calculating needs of industrial nations, though these machines required multitudes of often unnamed human computers to operate in a systematic way.

# TWO

# The First Electronic Computers

## THE ABC COMPUTER

When John Vincent Atanasoff (1903–1995) turned ten years old, his family moved into a new house at the phosphate mine where his father worked as an electrical engineer. For the first time, the young boy lived in a home with electrical lights. The family also bought their first automobile that year, at a time when most Americans could not afford automobiles. During that same year of wonders, his father also bought a new Dietzgen slide rule, but found that he really did not need such a sophisticated mathematical tool when he spent most of his time arranging for the repair of mining equipment and facilities. Atanasoff picked up the book of instructions and taught himself how to use the slide rule, performing addition, subtraction, multiplication, division, and then more sophisticated mathematical functions, such as logarithms and trigonometric functions. His father's book on college algebra helped him understand this new world of more advanced mathematics.

An avid reader, Atanasoff also found himself reading an old mathematics book of his mother's, where he was introduced to the idea of different bases for numbers. We use base 10, with digits from 0 through 9. The Babylonians used base 60 and the Mayans base 20. The idea of number bases was at that time mostly a mathematical curiosity with little practical utility, but base 2 was to prove useful in computing. While still in high school, Atanasoff decided that he wanted to be a theoretical physicist, though when he attended the University of Florida, he found no classes in theoretical physics. The courses that electrical engineering offered were the most challenging, so he graduated in 1925 as an electrical engineer, like his father, and moved from the humid South to the dry plains of Iowa to start graduate studies in mathematics at Iowa State College, now known as Iowa State University. He earned a master's degree in mathematics from Iowa State and a doctorate in physics from the University of Wisconsin in 1930. Then he returned to Iowa State College as an assistant professor of mathematics.

Atanasoff was frustrated by how long it took to calculate the results of a large number of calculations. Desktop mechanical calculators, manufactured by companies such as Monroe, Marchant, and others and powered by a hand crank, were used in these efforts, but mathematicians found them tedious to use. It might take weeks of work to solve a large set of equations. Desiring a machine that could solve partial differential equations, Atanasoff surveyed technology to see what might be available. An IBM tabulating machine in the statistics department used mechanical counters and intrigued Atanasoff, but he found that it really only added up categories of information on punched cards and could not solve equations. Atanasoff was the first to apply the word *analog* to machines that used mechanical counters. Digital computers store their numbers as distinct digits, with sharp boundaries between each number, whereas the numbers in analog computers have values that can smoothly become other values.

Other analog mechanical computers included slide rules, the differential analyzer built by Vannevar E. Bush (1890–1974) at the Massachusetts Institute of Technology (MIT), machines that used Fourier analysis, and antiaircraft fire directors. The last were machines that calculated how far to lead the antiaircraft gun, based on the height and speed of the target, so that the shells would intersect with the aircraft. Atanasoff realized that he wanted to use a digital computer (though he did not coin the word *digital*). Digital computers that already existed included the Chinese abacus, some bookkeeping machines built by Burroughs, and the desktop mechanical calculators that he was already familiar with. Atanasoff thought of buying thirty or so Monroe mechanical calculators, arranging them in a line, and driving them simultaneously by a common shaft. The problem with this was that each calculation would have to be recorded by hand and inputted by hand on each machine, leading to a high risk of mistakes. A single mistake would render the calculation inaccurate.

Since what he wanted did not exist, Atanasoff decided that he would have to invent a computer, though he did not approach the project with enthusiasm. He was teaching both mathematics and physics and was father to a family with three young children. He began to theoretically design such a computer and decided to use vacuum tubes, which were heavily used in radio technology but thought too unreliable by many experts for electronic applications. Vacuum tubes were electrical devices that could amplify electrical signals and act as switches. To make the computer more reliable, Atanasoff elected to use binary digits. He was unusual in that he was familiar with base 2 when many other scientists and engineers were not. Atanasoff also wanted his computer to have an electronic memory as it made calculations (the same idea that Babbage called a *store*) and coined the term *memory* in this context.

Frustrated with the lack of progress in his theorizing, Atanasoff took a long drive during the winter of 1937, traveling at excessive speeds, and eventually arrived in Illinois. After a couple of drinks at a roadhouse, the solutions to his problems became clear. He would use condensers (capacitors) for his memory, and to keep them from gradually losing the bit values put in the condensers, he would periodically pass electricity through the condensers to refresh them. He called this *jogging*, and it is the same principle of refreshing used in modern computer memory chips today. He also decided to create logic circuits to perform addition and subtraction, instead of using enumeration, as mechanical computers did. He drove home much slower, relieved to have broken through the mental barriers and solved so many problems.

More months of theoretical introspection followed as Atanasoff expanded on the ideas that jelled during his roadhouse visit. In 1939, he received funding from the college to build a prototype and hire an assistant. A brilliant young electrical engineering graduate student was recommended to him, and Atanasoff was fortunate to hire Clifford E. Berry (1918–1963). Working in a basement next to a student workshop, the two men carefully built each component and tested it thoroughly before moving on. They found vacuum tubes were expensive, used a lot of space, generated too much heat, consumed too much power, and often failed.

Atanasoff and Berry completed a working prototype before the end of 1939, and though it could only add and subtract binary numbers, the machine presaged the future. The computer was digital; used vacuum tubes, binary numbers, logic circuits, and refreshing memory; and had a rotating drum containing condensers to serve as memory. Rotating drum memory became popular for a couple of decades but is no longer used. The computer used a mechanical clock driven by an electrical motor. The clock in a computer is like the metronome for a music student in that it keeps everything synchronized. Atanasoff wrote a manuscript in 1940 describing the theory of his computer, his plans for the future, and how the computer would solve large systems of linear algebraic equations. He used the manuscript to obtain more funding, and the Research Corporation awarded him a grant of $5,330. While the two men worked on creating a complete machine, a patent attorney was contacted. Iowa State College and Atanasoff agreed to share the patent, but uncertainty by the patent attorney over what documentation would be required for such a new device led to a delay in the patent application.

In December 1940, Atanasoff introduced himself to John W. Mauchly (1907–1980) at a meeting of the American Association for the Advancement of Science (AAAS) in Philadelphia because Mauchly had presented a paper on a harmonic analog analyzer that he had developed. The analog analyzer performed Fourier transforms. Atanasoff was excited to meet someone also interested in computing and invited Mauchly to visit him in Iowa. Mauchly came during the summer of 1941 and stayed for five days, reading Atanasoff's research manuscript, examining the partially completed computer, and talking with Atanasoff and Berry about the invention. He also took notes on the manuscript, though Atanasoff would not let him have a copy of the manuscript because the patent application process was not completed.

So that human operators could work with their normal number system, Atanasoff and Berry added a device to the computer to convert to and from decimal (base 10) and binary (base 2) numbers. The computer could solve twenty-nine equations with twenty-nine unknowns and used punched cards, like IBM tabulating machines, to hold numbers beyond the capacity of their memory drum. Atanasoff and Berry completed their computer in early 1942, and it worked, though errors due to flaws in their punch card stock occasionally cropped up.

The Japanese attack on Pearl Harbor on December 7, 1941, changed everything for the two men. In June 1942, Berry moved to California to work at the Consolidated Engineering Corporation. In September 1942, Atanasoff moved to the Naval Ordnance Laboratory in Maryland, where he directed work on acoustics for the navy for the rest of the war. Neither man ever returned to their work at Iowa State College. Sadly, the college failed to recognize the jewel in the basement, and the patent application was never actually submitted by the patent attorney. Since Atanasoff and Berry were gone, their computer was dismantled.

Last known photograph of the ABC computer before being dismantled by officials at Iowa State. (Image Courtesy of the Charles Babbage Institute Archives, University of Minnesota Libraries, Minneapolis)

Atanasoff occasionally met Mauchly during the war, and Mauchly told him that he was working on a computer based on completely different principles than Atanasoff's. We now know that was not true, since many of Atanasoff's ideas were incorporated into Mauchly's work. After the war, Atanasoff continued to work for the military in their Bureau of Ordnance, including an effort to build a computer for the bureau. The computer project was canceled after a short time, and Atanasoff moved on to other projects. In 1949, he served as chief scientist for the U.S. Army Field Forces, and in 1952, he founded his own company, the Ordnance Engineering Corporation. In 1956, he sold the company to Aerojet General Corporation, and after a time as an executive with Aerojet, Atanasoff retired in 1961.

Mauchly and his colleague, J. Presper Eckert (1919–1995), filed for a patent on their ENIAC (Electronic Numerical Integrator and Computer), in 1947, and for many years, historians considered the ENIAC to be the first electronic digital computer. While ABC was effectively a prototype, the ENIAC was a fully functioning computer doing useful work. The ENIAC patent was later owned by the Sperry Rand Corporation, and lawsuits eventually began when Sperry Rand asked for royalties from other computer manufacturers. Lawyers contacted Atanasoff and asked for his help, prompting Atanasoff to examine the ENIAC patent. He was surprised to find many of his own ideas in the patent and participated in an epic legal battle to overturn the Mauchly and Eckert patent. Berry did not participate because he had apparently committed suicide in 1963 for unknown reasons. To honor Berry's contributions to their joint effort, Atanasoff started to refer to their computer as the Atanasoff-Berry Computer, or the ABC. In a 1973 federal court decision, the Mauchly and Eckert patent was set aside and the ABC declared the first electronic digital computer. The judge who handed down this decision did not know of efforts in Germany with the Zuse machines and some efforts in Britain that could also lay claim to being the first electronic digital computers.

Though Atanasoff received no royalties from inventing the computer, he was showered with awards after the 1970s. The communist nation of Bulgaria, proud of a man who had an obvious Bulgarian name and was the son of an emigrant from Bulgaria, awarded Atanasoff its highest scientific honor. Among other awards and honorary degrees that graced Atanasoff's later years were the Pioneer Medal from the Institute of Electrical and Electronics Engineers (IEEE) Computer Society in 1984, an IEEE Electrical Engineering Milestone in 1990, and a Medal of Technology in 1990 from the U.S. Department of Commerce.

## CODEBREAKING WITH BOMBES AND COLOSSI

Sailors have always had reason to fear the sea. Stories of shipwrecks and ships lost at sea, victims of rocks or harsh weather, are common fare. During World War II, Allied sailors in the Atlantic experienced an additional fear, as thousands of them lost their lives to torpedoes from German U-boats. Unknown to these men, as they waited in the darkness and wondered whether they would live, some of the first computers were helping them to survive. These computers remained unknown to historians for decades after the end of the war.

World War II was a war of science and technology as much as it was a struggle between fighting men. Part of that struggle involved codebreaking. Competent military commanders have always tried to keep their plans secret from their enemies, which led to the rise of codes to conceal the contents of written messages. With the coming of radio communications in the twentieth century, codes were applied to radio traffic to conceal their content from enemy eavesdroppers. By World War II, codes had become so sophisticated that mechanical help was required to encode and decode messages.

The German Enigma encoding machine invented in 1919. This electromechanical encryption device spurred the invention of the secret bombe and Colossus computers by the British during World War II. (National Cryptologic Museum/National Security Agency)

The Enigma encoding machine was patented in 1919 by a German company for commercial use: concealing messages from possible business competitors. In 1926, between the two World Wars, the German navy adopted the Enigma machine to encode its radio traffic. Other branches of the Germany military and German government departments followed suit in the next decade. The Dutch military also purchased Enigma machines and began to use them in 1931. In 1943, the Germans shipped 500 Enigma machines to the Japanese for use in German-Japanese communications. A limited number of Enigma machines were also used by Italy and other German allies.

The electromechanical Enigma looked like a portable typewriter in a small wood box. A lamp board above the keyboard contained the twenty-six letters of the alphabet. A set of three rotors above the lamp board could each be rotated to twenty-six different positions. The operator set the rotors to a daily prearranged setting; typed in a message, which sent voltage through the machine, letter by letter; and the encoded or decoded letters appeared on the lamp board. After encoding a message and copying the resulting letters down, the operator then sent the message via telegraph or via radio using Morse code. The message was transmitted in the clear but read like gibberish. Because the Enigma encryption scheme was symmetric, the receiver of an encoded message only needed to have the same correct rotor settings to decode the message. As an operator typed the encrypted gibberish into an Enigma machine, the lamps on the lamp board lit up with each of the decoded letters in correct order.

The German military added a plugboard to its Enigma machines that allowed up to six pairs of letters to be interchanged. The plugboard, combined with the three rotors, resulted in many millions of possible combinations. Certain that no one could break messages sent by their machines, the Germans allowed the continued sale of commercial Enigma machines. The Poles detected that the Germans were using this new machine and bought one for themselves, added the military plugboard to it, and tried to figure out a way to break the daily code settings. They failed until the mathematician Marian Rejewski (1905–1980) began working on the problem and developed a decryption technique that often worked to decode a set of messages. The Poles also built an electromechanical machine to help them in their work. The ticking sound of the machine prompted them to call it a *bomba*, the Polish word for "bomb."

When Adolf Hitler decided to invade Poland, his intentions became obvious to the Poles from decoded radio intercepts. Six weeks before the war began, the Poles called a meeting with their allies and revealed to the British and French the extent of their success. The Poles gave the British a copy of the Enigma machine, the plans for their bomba, and a copy of the statistics that they had gathered from intercepted message traffic that allowed them to more easily break new Enigma traffic. Remarkably, even after Poland and France both fell to the German armies within a year of this important meeting, the Polish breakthrough remained a secret.

The British realized the jewel that had been handed to them and created an organization to exploit it. They built large radio receivers to pick up radio messages bouncing off the atmosphere from deep inside Nazi-occupied Europe. These messages were transcribed and carried to a government-owned estate outside of London called Bletchley Park. The British recruited the best and brightest to work in spartan conditions at Bletchley Park to decode the Enigma traffic. Women from the auxiliaries of the British armed forces formed the backbone of the effort. They hunched over radio receivers and typewriters as they engaged in the detailed work of indexing thousands of radio intercepts a day and keeping the flow of paper going.

The Germans usually changed their Enigma rotor and plugboard settings every day, using printed code books so that distant military units knew the settings for each day. They also made the rotors removable. A set of five rotors (eight on naval models) was supplied with each machine, with only three being used each day. Different branches of the German military used different codes, so the navy, weather service, different commands of the army, and so on had different codes for a given day. This meant that the wizards at Bletchley Park had to break the code for a given day and for a given service, which was a never-ending effort. Mistakes by German operators and standard formats for certain types of messages helped the British codebreakers. At times, the British even planted information, hoping to create a situation where the codebreakers might already know the content of an encrypted message.

The brilliant British mathematician Alan Turing (1912–1954) served as a leading codebreaker at Bletchley Park. Turing's skills in mathematics were recognized at an early age, though he was not a good student at the boarding schools he attended. After a couple of failures to gain admission to college, he was accepted at King's College at Cambridge University and graduated with a master's degree in mathematics in 1934. His 1936 paper, "On Computable Numbers," contained an argument about what kinds of problems are computable and proposed a theoretical computer that became known as the *Turing machine*. After a couple of years studying in the United States at Princeton University, Turing returned to Cambridge and became involved in theorizing efforts on how to build a computer. He and his colleagues decided to use binary numbers and Boolean algebra. A reflection of the type of occurrence common in the history of technology, where multiple people separately make the same leap of innovation, Turing, Atanasoff, and other computer pioneers all concluded that binary numbers were the solution to making the electronics of their computers simpler.

When World War II broke out, Turing was recruited into Bletchley Park. Turing and the British were completely unaware of the obscure efforts of Atanasoff when Turing and other codebreakers redesigned the bomba, which the British called a *bombe*. The first bombe was built in 1940 at the British Tabulating Machine Company in Letchworth. The British bombes were essentially electromechanical reproductions of 12 Enigma machines, with each emulating the rotor settings. These bombes did not break the code, but they excluded possibilities, leaving the remaining possibilities to be broken by hand. The noisy bombes broke frequently and required almost constant repair. The British eventually shared their secrets with the United States, and the Americans built their own versions of the bombes. Faster bombes that emulated up to 36 Enigma machines and weighed over a ton were built by both the British and the Americans in 1943. Over 100 bombes in total were built during the war.

Besides the Enigma machines, the German army also started to use Lorenz SZ42 cipher machines during the war, especially for high-level communications between Berlin and distant armies. These machines encrypted their teleprinter traffic through an encryption system invented by Gilbert Vernam (1890–1960), an American, during World War I. The Lorenz machine was superior to the Enigma machine in that it both encrypted and transmitted its teleprinter traffic, and it automatically received and decrypted the messages at the receiving end. This system relied on a set of randomly created characters that were interspersed with the cleartext before encryption. At the receiving end, the same set of randomly created characters were used to decrypt the message. If a truly random set of characters was created and then copied for use by the sender and receiver, the system was theoretically unbreakable. The Lorenz engineers realized the difficulty of distributing such random sets, so they built into the machine the ability to create a pseudorandom character set. These automatic pseudorandom characters sets weakened the strength of the cipher. Whereas Enigma machines were mainly used for tactical purposes, the Lorenz machines were used to transmit longer communiqués of strategic value, such as order of battle information, supply reports, and military planning discussions.

The British detected radio traffic from the Lorenz machines in 1940 and code-named the unknown messages "Fish." Codebreakers at Bletchley Park figured out how to break the code, but the process took so long that the decrypted messages might be weeks old and the intelligence grown stale. The codebreakers built a machine called the Heath Robinson (named after a cartoonist known for his drawings of fanciful machines) that read two paper tapes, slowly searching for a match. Tommy Flowers (1905–1998), a mechanical engineer and telephone exchange expert at the Post Office Research Labs at Dollis Hill, took the concept from there and built the Colossus. Users programmed the digital Colossus via its plugboard and switches. The Colossus was also an example of an early computing effort to manipulate symbols in the form of letters rather than just serve as number crunchers. The Colossus was a special-purpose machine, effectively a sophisticated signal processor, not a general-purpose computer that that could be reprogrammed to do other tasks.

The original Colossus machine used 1,500 thermionic valves (vacuum tubes). A paper tape with an encrypted message on it was fed into the machine at over 1,000 characters per second. The British built a total of ten Colossi during the war, eight of which were the more advanced Mark II version with 2,400 vacuum tubes. The Mark II machines ran five processing units in parallel, reaching an effective total speed of 25,000 characters per second. The Colossi decrypted a total of 63 million German characters before the end of the war. By 1943, Bletchley Park was regularly reading Fish traffic after a delay of only a few days. What is remarkable about the effort to break the Lorenz machines is that the British succeeded without ever capturing or learning any details about the actual machines.

The British and Americans used the term *Ultra* to describe decrypts that came from Bletchley Park. Keeping Ultra secret was considered as important as keeping the Manhattan Project to build the atomic bomb secret, and Ultra intercepts significantly helped the Allies win the war against Germany. The secrets of Bletchley Park were not released to the public after World War II. Eight of the Colossi were immediately destroyed, and the last two were destroyed in about 1960 and their blueprints burned. In the 1970s, details about Bletchley Park slowly became known. As part of an effort to reclaim this lost history, a new Colossus was built at a museum in England, relying on memories of surviving engineers, what few pictures and diagrams were not destroyed, and information that the Americans had retained about the machines.

Turing's contribution to the success of Bletchley Park was also kept a secret. After the war, Turing continued to work on computers and spent time working on an Automatic Computing Engine (ACE) at the National Physical Laboratory before moving on to serve as deputy director of the Royal Society Computing Laboratory at Manchester University. Turing published a famous paper on artificial intelligence in 1950 that proposed a test for determining whether a computer was intelligent. After being convicted of a crime related to his homosexual behavior and serving his sentence, Turing killed himself by eating an apple laced with cyanide.

## THE ZUSE COMPUTERS

Konrad Zuse (1910–1995) was born in Berlin and had dreams of designing great cities and moon rockets. He attended the Technische Hochschule in Berlin and became a civil engineer in 1935. While in school, Zuse conceived of a machine to automatically solve systems of linear algebraic equations, and in 1936, he began to design a mechanical computer that he later called the *Z1*. He chose to use binary, coming to this idea a year before Atanasoff in Iowa. During the day, Zuse worked as an aircraft engineer conducting stress analysis for the Henschel aircraft company; at night and on weekends, he worked on his computer, funding the effort out of his own salary. A friend, Helmut Schreyer (1912–1984), helped Zuse in his work.

Zuse chose to leave Henschel to devote more time to his computer project, and in 1938, he finished the Z1 in the living room of his parents' home. Instructions were fed into the machine by using old movie film as punched tape. Memory was maintained via pins in small slots cut into sheet metal. The machine used a floating point format to represent complex numbers, a significant innovation compared to the effort by Atanasoff. The arithmetic unit of the machine could only work for a few minutes before errors cropped up. The patent claim that Zuse filed with the U.S. Patent Office was rejected because of insufficient detail.

Zuse and Schreyer did not give up. Instead, they began to work on their next computer, the Z2. This computer was electromechanical, using secondhand relays. They wanted to use vacuum tubes, but the number required was beyond their financial means. The coming of World War II resulted in Zuse being drafted into the army, and a year passed before he was discharged to return to his old job with Henschel. Zuse completed the Z2 on his own and demonstrated it to the German Aeronautical Research Institute. The German government agreed to fund his effort to build a Z3 but did not give him sufficient support to do more than continue to work at his home. The Z3 was finished before the end of 1941 and contained 2,600 relays, but it still used mechanical memory. The Z3 was faster in multiplying numbers than the electromechanical Harvard Mark I, but much slower than the all-electronic ENIAC. An air raid destroyed the Z3 in 1944.

Given more support, Zuse founded a company, Zuse Apparatebau, in 1942 and started to build the Z4. In the meantime, his company built several special-purpose calculators for Henschel to calculate wing and rudder surfaces for aircraft and flying bombs. The coming end of the war caused Zuse to flee to the small town of Hinterstein in Bavaria, where he hid the dismantled Z4 computer in the basement of a farmhouse. The Z4 was eventually retrieved and completed. It was set up for use at the Federal Polytechnical Institute in Zurich, Switzerland, in 1950 and then moved five years later to the French Aerodynamic Research Institution, where it was used until 1960.

During the chaos after the end of World War II, Zuse found the time and mental focus to create one of the first computer programming languages, which he called *Plankalkul*. Historians have been impressed by a language that used variables, conditional and looping statements, and procedures. Zuse also formed a company after the war, Zuse Kommandit Gesellschaft (Zuse KG), which continued to make mechanical computers for the European market. In 1958, the Z22 became the first Zuse computer based on vacuum tubes. Zuse KG eventually became part of the large German firm Siemens. While Zuse's efforts were impressive from a historical point of view, his machines did not influence the development of later computers. In a sense, his efforts were a historical diversion, a result of being isolated from the dynamic technological innovation going on in the United States and Britain.

## THE HARVARD MARK I

Another computer pioneer, Howard Hathaway Aiken (1900–1973), was born in Hoboken, New Jersey, but grew up in Indianapolis, Indiana. In high school, he began working for the Indianapolis Light and Heat Company. He continued to work in the electrical utility industry while in college at the University of Wisconsin–Madison. Graduating as an electrical engineer in 1923, he remained in the same field of work for another nine years, when he decided that he had picked the wrong career. He wanted to be a physicist, so he enrolled for a year at the University of Chicago to study mathematics and physics before moving on to Harvard University. He obtained his master's degree in 1937 and a doctorate in 1939.

Like other computer pioneers, Aiken wanted to solve large systems of equations, which led him to think about building a computer. He had begun work on a calculating machine when a technician took him into an attic at Harvard and showed him a piece of the Babbage calculating engine that had been donated to Harvard by Babbage's son in 1886. Aiken was fascinated and studied the work of Charles Babbage, coming to view himself as Babbage's successor. Aiken approached Thomas J. Watson Sr. (1874–1956), the president of IBM, for funding. Watson generously supported Aiken's effort, and IBM engineers did most of the design work as well as actually building the machine. Unlike the mechanical differential analyzer built by Vannevar E. Bush at MIT, Aiken's proposed electromechanical machine was designed to perform all kinds of mathematical operations, not just differential equations.

After World War II started, Aiken served in the navy before being asked to return to Harvard to direct the U.S. Navy Computing Project. After spending half a million dollars, the Mark I was completed in 1943. IBM had paid for two-thirds of the cost, while the navy picked up the rest of the total cost. The long, narrow machine stretched fifty-one feet from side to side, stood eight feet high, and was only two feet deep. Weighing five tons, the Mark I included three million wire connections. The heart of the machine was seventy-two IBM mechanical rotating registers. Unlike the British bombes, the machine was as quiet as a few typewriters. Programming the machines required adjusting 1,400 switches, using a paper tape to feed in instructions and punched card readers to input data.

Grace Hopper (1906–1992) worked closely with Aiken on the Mark I computer and later became an important figure in the development of programming languages. Hopper, born Grace Brewster Murray in New York City, attended private schools for girls, then the women-only Vassar College in Poughkeepsie, New York, before attending Yale University in 1928. In 1934, she graduated with a doctorate in mathematics and mathematical physics from Yale. Social attitudes against women in the workplace limited opportunities for women at that time, and she turned to teaching at Vassar. She married in 1930, taking the last name of Hopper, though the childless marriage ended in divorce in 1945.

In 1943, Hopper joined the navy and was assigned to the navy's computer project at Harvard a year later. Aiken assigned her to read the writings of Babbage and to write the manual for the Mark I, which led to her programming the Mark I. Aiken and Hopper published joint scholarly articles on their efforts, establishing her reputation as a programmer. Aiken went on to build the Mark II for the U.S. Navy, based entirely on electromagnetic relays, and several more Mark computers in the 1950s, each using ever more advanced technology. While Aiken's efforts led to useful computers, and the Harvard Mark I was one of the most impressive electromechanical computers ever built, the modern computer, with stored programs and being entirely based on electronics, traces its lineage from the ENIAC computer.

## THE ENIAC

Mauchly and Eckert formed a team that managed to bring the electronic computer out of government laboratories and into the commercial world of data processing, where they launched a revolution. John W. Mauchly was born in Cincinnati, Ohio, and grew up near Washington, DC, where his father was a physicist at the Carnegie Institute. He enrolled at Johns Hopkins University in Baltimore, Maryland, in 1925, and after two years as an undergraduate, he applied to enroll directly in a doctoral program in physics. He earned his doctorate in 1932, emphasizing the study of molecular spectroscopy, and joined the faculty of Ursinus College in Collegeville, Pennsylvania.

Like most scientists, Mauchly was frustrated by how long it took to solve large systems of equations, especially since his research focus had turned to studying weather systems. He hired a group of graduate students in mathematics to use mechanical calculators to solve the large number of equations he needed to understand the statistics behind the effects of a solar flare on the weather. That tedious exercise prompted him to build a machine to solve Fourier transforms and led him to present a paper on it at the AAAS Philadelphia meeting, where he met John A. Atanasoff. After he returned from examining Atanasoff's work in Iowa, Mauchly took an advanced course in electronics at the Moore School of Electrical Engineering at the University of Pennsylvania and was asked to remain on the faculty of the school. Mauchly met Eckert soon after arriving at the Moore School. A native of Philadelphia, J. Presper Eckert had completed his bachelor's degree at the Moore School in 1941 and had remained at the school for his graduate studies.

The Moore School signed a research contract with the Ballistics Research Laboratory (BRL) of the U.S. Army, and in an August 1942 memorandum, Mauchly proposed that the school build a high-speed calculator using vacuum tubes for the war effort. In 1943, the army granted funds to build the Electronic Numerical Integrator and Computer (ENIAC) to create artillery ballistic tables. Eckert served as chief engineer of a team of fifty engineers and technical staff on the ENIAC project.

Completed in 1945, the ENIAC consisted of forty 8-foot-high cabinets, almost 18,000 vacuum tubes, and many miles of wiring; it weighed thirty tons. To minimize the high failure rate of vacuum tubes, Eckert ran the tubes at a lower voltage than they were designed to handle. A plugboard was used to program the computer. A ballistic trajectory calculation that took a human mathematician twenty hours to solve was completed by the ENIAC in thirty seconds. Oddly enough, the ENIAC used decimal numbers instead of binary numbers.

The American ENIAC machine was completed two years to the month after the first British Colossus. If the ABC was a rowboat, the ENIAC was

The ENIAC computer. (Image Courtesy of the Charles Babbage Institute Archives, University of Minnesota Libraries, Minneapolis)

a three-masted sailing ship ready to carry cargo, though several of the key original innovations probably came from the work of Atanasoff and Berry. While building the ENIAC, Mauchly and Eckert developed the idea of the stored program for their next computer project, where data and program code resided together in memory. This concept allowed computers to be programmed dynamically so that the actual electronics or plugboards did not have to be changed with every program.

The noted mathematician John von Neumann (1903–1957) became involved with the ENIAC project in 1944 after a chance encounter with an army liaison officer at a railroad station. Johann (later anglicized to John) von Neumann was born in Budapest, Hungary, to a Jewish family; his father was a banker. His family recognized his extraordinary intelligence as a child and hired a private tutor to supplement his education. When he received his doctorate in mathematics from the University of Budapest in 1926, von Neumann was only twenty-two years old and already publishing mathematical articles. In 1930, von Neumann moved to the United States, where was a visiting lecturer at Princeton University before becoming a full professor and an original member of Princeton's Institute for Advanced Study only three years later. His prowess in mathematics and numerous original contributions made him a leading theorist in game theory and set theory. During World War II, von Neumann worked on the Manhattan Project to build the atomic bomb and also lent his wide expertise as a consultant on other defense projects.

After becoming involved in the ENIAC project, von Neumann expanded on the concept of stored programs and laid the theoretical foundations of all modern computers in a 1945 report and through later work. His ideas came to be known as the "von Neumann Architecture," and von Neumann is often called the "father of computers." The center of the architecture is the fetch-decode-execute repeating cycle, where instructions are fetched from memory and then decoded and executed in a processor. The results of the executed instruction change data that are also in memory. After the war, von Neumann went back to Princeton and persuaded the Institute for Advanced Study to build its own pioneering digital computer, the IAS (derived from the initials of the institute), which he designed.

Along with honorary doctoral degrees, von Neumann was elected to the National Academy of Sciences in 1937 and received several national honors for his defense work. He died of cancer in 1957, perhaps contracted during his work on the Manhattan Project. Von Neumann was a gregarious man with sophisticated tastes, a command of four languages, a prodigious memory, and an amazing ability to perform calculations in his head.

Eckert and Mauchly deserve equal credit with von Neumann for their innovations, though von Neumann's elaboration of their initial ideas and his considerable prestige lent credibility to the budding movement to build electronic digital computers. The Electronic Discrete Variable Arithmetic Computer (EDVAC) was designed by the Moore School as the successor to the ENIAC and was to be the first stored-program computer, using binary numbers instead of decimal numbers, but the departure of Mauchly and Eckert in 1946 to form their own commercial venture delayed the completion of the EDVAC until 1952.

## THE MANCHESTER MARK I AND THE EDSAC

The Moore School decided to hold an eight-week summer school on the Theory and Techniques for the Design of Electronic Digital Computers during the hot months of 1946, instructing invited scientists and engineers in the new art of computers. This onetime event, later known as the Moore

School Lectures, effectively spread the knowledge created at the Moore School and prompted other digital computer efforts. Multiple mathematicians and engineers from Britain visited the Moore School or attended the lectures. Some of these British scientists and engineers had worked on the bombes and Colossi of Bletchley Park and so had additional knowledge that they were required to keep secret. The others had usually worked on the war effort in some capacity. F. C. Williams (1911–1977), on the engineering faculty at the University of Manchester in Britain, visited the Moore School that summer and noted the work on creating memory storage via mercury acoustic delay lines or by using cathode-ray tubes (CRTs). Williams returned to the Telecommunications Research Establishment (TRE), where he had worked on radar and Identification Friend or Foe (IFF) aircraft systems during the war. Working with his colleague, Tom Kilburn (1921–2001), Williams developed a way to store binary bits inside CRTs, which became known as Williams tubes.

CRTs work by shooting an electron beam at a screen covered with phosphor dots. The electrons in the beam react with the phosphor atoms to release photons as light. This effect persists for a fraction of a second and will disappear unless the electron beam refreshes the phosphor dot. The Williams tubes took advantage of this delay to store the values of bits on the surface of the CRT. By changing the intensity of the electron beam from a writing mode to a sensing mode, the beam could sense whether a bit was displayed on the screen. Because the effect of the electron beam on the phosphor atoms rapidly deteriorated, memory had to be regularly refreshed by sensing the bit values and then rewriting them.

Williams and Kilburn both moved to Manchester University, where a group of mathematicians and engineers who had worked on radar research or at Bletchley Park during the war came together to work at the Royal Society Computer Laboratory. Alan Turing also joined them for a time. In 1948, the Manchester team succeeded in running a simple stored program on their prototype machine. Their machine had only three CRTs, one of which stored thirty-two words, each thirty-two bits long, with the other two CRTs being used as an accumulator and to hold the instruction under execution. After completing the prototype, later called the Manchester Mark I, based on Williams tube electrostatic memory units and the use of a rotating drum to magnetically store data, the Manchester team contracted with an outside firm to build the first production model. The production model was also called the Mark I and was built by a local electronics firm, Ferranti. The first Ferranti Mark I, delivered to Manchester University in spring of 1951, actually beat the more famous UNIVAC as the first commercially produced computer to be delivered to a customer. The Ferranti Mark I contained 4,000 vacuum tubes, 2,500 capacitors, and 12,000 resisters. A team at Manchester University also continued to develop more sophisticated computers, laying the foundation for a commercial computer industry in England.

The British mathematician Maurice V. Wilkes (1913–2010) also attended the Moore School Lectures, and on the voyage home to Cambridge University, he began to design his own computer. His goal was simply to create a stored-program computer, though Wilkes correctly perceived that the main technical problem was how to store memory. A Cambridge team led by Wilkes developed a way to temporarily store an electrical impulse in a tube of mercury, the mercury delay line. The Electronic Delay Storage Automatic Calculator (EDSAC) performed its first calculation in 1949 and was a functional stored-program computer. The Cambridge team rapidly innovated in programming, developing a symbolic notation system and even the idea of subroutines, which are segments of reusable code. Wilkes and two colleagues wrote the first book on programming, *The Preparation of Programs for an Electronic Digital Computer*, in 1951. This textbook strongly influenced other projects, both in Britain and back in the United States.

## THE FIRST PROGRAM

What is a *program*? Early computers all ran programs in that the hardware performed a sequence of mathematical operations. What we now think of as a computer program (or *stored program*) is the result of the work of von Neumann and the other innovators on the ENIAC and the EDVAC. A year after its creation, the ENIAC was disassembled and moved to its permanent home at the Aberdeen Proving Ground, an army base in Maryland. When it was reassembled in 1947, the newer ideas of a stored-program computer were used to modify the machine into something very similar to what we think of as a modern computer. The improved machine was even able to run programs directly from punched cards, and the ENIAC programs included conditional jump instructions and other innovations that define modern programs.

Von Neumann and other nuclear weapons researchers at the Los Alamos National Laboratory in New Mexico wanted to use the modified ENIAC to simulate the movement of free neutrons during a nuclear explosion. Many physics equations do not have a single solution, and the equations must be run numerous times to gather statistical data. This is called a *Monte Carlo simulation*, named after the famous casino in Monaco, and was very difficult to do. Before electronic computers, dozens or hundreds of human computers were used instead. Von Neumann and his team developed a program for the ENIAC, and the first successful series of runs was completed in 1948. The atomic bomb had already been invented—and used on Hiroshima and Nagasaki—but this Monte Carlo simulation helped scientists to better understand how to make the fission reaction more efficient.

Two of the important figures working on the ENIAC and the Monte Carlo program were Klara von Neumann (1911–1963), John's wife, and Adele Goldstine (1920–1964), the wife of Herman Goldstine (1913–2004). Herman was the army liaison officer who had introduced von Neumann to the ENIAC project, and he continued to work on the ENIAC and later joined von Neumann at the Institute for Advanced Study. Adele was a brilliant mathematician in her own right and an important innovator. Among her accomplishments, Adele wrote the manual that described how the ENIAC operated and how to program it. True solitary inventors are rare in the history of technology, especially as technology has become ever more complex. Teams were the real innovators, with each person contributing their part, yet we are drawn to stories of individuals rather than the more diffuse credit found with teams and entire industries. Women's roles have often been neglected in historical accounts, partially because of this bias against team-oriented narratives.

## THE UNIVAC

In 1946, Mauchly and Eckert managed to persuade the U.S. Census Bureau to purchase a computer called the UNIVAC (UNIVersal Automatic Computer). The bureau contracted to spend $300,000, though Mauchly and Eckert thought that developing the machine would cost $400,000. They hoped to recover their financial loss by selling more computers. Hiring engineers and technicians, they worked long hours as financial difficulties beset their Philadelphia-based company. A contract with Northrop Aircraft Corporation to build a small guidance computer called the BINAC (BINary Automatic Computer), for a guided missile, helped keep the company afloat. In 1948, the A.C. Nielsen market research firm and the Prudential Insurance Company also contracted to buy UNIVACs, though for substantially less than the U.S. Census Bureau paid. Desperate for money to finance their research and development activities, Mauchly and Eckert sold 40 percent of their corporation to the American Totalisator Company of Baltimore for $500,000 and secured additional loans to save the company.

By 1949, the Eckert-Mauchly Computer Corporation was doing well, with 134 employees and six orders for their UNIVAC, which was still under development. The death of the founder-president of American Totalisator brought financial calamity, as the subsequent management of American Totalisator called in its loans. Mauchly and Eckert tried to sell their company to IBM, but IBM declined the offer on the advice of its lawyers, fearing antitrust problems since IBM already controlled much of the mechanical calculator and tabulating machine market. Remington Rand agreed to buy the company as a wholly owned subsidiary, and Mauchly and Eckert became employees in their own company.

The UNIVAC contained 5,000 vacuum tubes that generated so much heat that the engineers and technicians worked in their shorts and undershirts during the hot Philadelphia summer of 1950. The UNIVAC was a true von Neumann machine, with 1,000 memory locations based on mercury delay lines, each capable of holding twelve digits or characters. Performing an addition operation took approximately half a millisecond, multiplication two milliseconds, and division four milliseconds. A variation on an electric typewriter served as a control console, and data were fed into the machine through magnetic tapes with nickel-coated bronze tapes, since early efforts with plastic tapes had failed. Early versions of the UNIVAC used air cooling, while later versions used a water-cooling system to dissipate the heat generated by the vacuum tubes. In 1949, Grace Murray Hopper was hired away from the Mark I project at Harvard as a senior mathematician to program the new machine. The first machine was finally delivered to the U.S. Census Bureau in the spring of 1951. Hopper remained with a succession of computer companies until the mid-1960s, when the navy reactivated her and kept her employed until 1986.

As a publicity effort, Remington Rand proposed to the CBS television network that it use a UNIVAC to predict the winner of the November 1952 presidential election. Mauchly worked for months with a statistician to create a program that used returns from the 1944 and 1948 elections to predict the results in key states from early votes. Early on the election

The UNIVAC computer. (Image Courtesy of the Charles Babbage Institute Archives, University of Minnesota Libraries, Minneapolis)

night, the UNIVAC program predicted such an overwhelming landslide for Dwight D. Eisenhower that the network executives refused to broadcast the results and compelled the programmers to change the program to make the race closer. As the true results came in, showing a crushing defeat for the Democrats, the network admitted that the first results had been correct.

Remington Rand built forty-six UNIVACs for government and business organizations and became the world's leading supplier of data processing computers. Remington Rand merged with Sperry Gyroscope in 1955 to become Sperry Rand, which later merged with Burroughs Corporation in 1986 to form Unisys, briefly the second-largest computer corporation in the world. Mauchly left Sperry Rand to form a consulting firm in 1959, while Eckert remained with the company for the remainder of his career.

Mauchly and Eckert were honored as the inventors of the electronic digital computer until 1973, when their patent was overturned in federal court and Atanasoff declared the inventor of the electronic digital computer prior to their efforts. In 1980, Mauchly and Eckert shared the IEEE Computer Society Pioneer Award. Even though their credit for the first computer was misplaced, Mauchly and Eckert did successfully develop the UNIVAC as the first commercial electronic computer in the United States. The original UNIVAC I ran for thirteen years before being donated to the Smithsonian Institution.

# THREE

# The Second Generation: From Vacuum Tubes to Transistors

## THE COLD WAR

World War II, the largest war in history, was fought on battlefields and in the laboratory. Never had a war been so dependent on research in science and technology. The inventions that poured forth, many of them created through the understanding of science, included radar, computers, jet airplanes, short-range missiles, and the atomic bomb. The Cold War developed in the late 1940s, a global ideological struggle between communism and centrally controlled markets, led by the Soviet Union, and democracy and free markets, led by the United States and its Western European allies. Actual combat encounters between the two superpowers proved to be rare—such as when reconnaissance aircraft were shot down—and were quickly concealed so as not to escalate the situation. Other nations instead served as proxies, fighting ideologically based surrogate wars.

Both antagonists realized how important science and technology were to their efforts, and funding levels for research into science and technology continued at unprecedented levels during the Cold War. The governments of the United States, the Soviet Union, and their allies recruited their best and brightest to serve in defense-related research and development. Scientists and engineers developed more advanced computers, computer networks, the internet, better medicines, better alloys, industrial ceramics, and technologies with no civilian use, like the neutron bomb.

In the United States, where most computer advances occurred, the military, space exploration, and other efforts by the federal government (especially the nuclear arms program) persistently challenged the emerging computer industry to make computers smaller, more reliable, faster, and more capable. Without this Cold War–induced spending, computer technology would likely have developed more slowly.

## PROJECT WHIRLWIND AND SAGE

During World War II, pilots trained in mechanical trainers before getting in real airplanes. Each airplane model required its own unique trainer. The navy wanted a single trainer that could be used for many models of airplanes and contracted with the Servomechanisms Laboratory at the Massachusetts Institute of Technology (MIT) to examine the feasibility of such a project. Jay W. Forrester (1918–2016), a bright young electrical engineering graduate student who had been raised on a cattle ranch in Nebraska, took charge of the project. The navy envisioned the new flight simulator using servomechanisms to move the simulator and control the flight instruments, all directed by an analog computer that could be reprogrammed to simulate each model of aircraft.

As Forrester worked on the project, he recognized that the simulator was a real-time system, where the computer must respond to real-world input within a set amount of time, and he decided that an analog computer would react too slowly to make the simulator work. When Forrester learned about the digital computer projects then going on, especially the ENIAC project, he determined that an electronic digital computer was necessary to make the simulator work. The navy agreed with Forrester and continued to fund the expanded project, which was named Project Whirlwind in 1946. Forrester and the navy also realized that the digital computer could be used in many other applications besides flight simulators. Whirlwind grew so much in the late 1940s that the navy became concerned about burgeoning costs and wanted to scale it back. Most flight simulators, like those created by Link Aviation Devices, remained analog machines through the 1950s.

The newly formed U.S. Air Force faced its own difficult problem; there was a fear that Soviet bombers flying in over the Arctic regions would drop atomic bombs on the United States during a war. Radars could detect the bombers, but how could they mesh command and control of all the radars and fighters into a single system to effectively counter the hypothetical Soviet attack? The air force turned to MIT, which established the Lincoln Laboratory to build the Semi-Automatic Ground Environment (SAGE) system. Needing a digital computer for SAGE, the air force took over funding the Whirlwind from the navy, and the Whirlwind engineers designed the FSQ-7 computer. The original purpose for Project Whirlwind, the flight simulator, was never built.

The air force contracted with International Business Machines (IBM) to manufacture the AN/FSQ-7 computers for SAGE, giving that company an important opportunity to continue to develop its expertise in electronic digital computers. At one point, one out of every five employees at IBM worked on the SAGE system. The AN/FSQ-7 computers were also the largest computers ever built, each containing over 49,000 vacuum tubes, weighing 250 tons, and occupying a three-story building. The SAGE software eventually totaled over a million lines of programming code, and at one point, over half of the programmers in the United States were working on this single project. By 1953, SAGE could simultaneously track forty-eight aircraft, and by 1963, the entire SAGE system of twenty-three direction centers, using twenty-four AN/FSQ-7 computers, was deployed, having cost about $8 billion. The SAGE system continued to function until 1983.

As with other digital computer projects, a key problem was how to store data in the computer. Forrester and the Whirlwind engineers looked at mercury delay lines, worked on using electrostatic storage tubes, and eventually perfected magnetic-core memory in 1953. This type of memory stored each bit in a tiny ring of ferrite material, about the size of a pinhead, which retained its on or off binary value magnetically, even if electrical power was turned off. This memory technology became the standard memory used in computers until integrated circuits offered a cheaper and faster alternative in the late 1960s.

The SAGE system was considered one of the great successes in early computer history, an example of systems engineering that combined numerous social and technological factors into a functioning whole. SAGE computers used telephone lines to communicate from computer to computer and from computer to radar systems, an early form of modems and networking. SAGE operators used video terminals to track information from the computers and radars. The light pen was invented as an input device for the operators. The largest real-time computer program written up to that time provided simultaneous access for SAGE operators. The Air Force even had ashtrays installed in the consoles to accommodate the long hours America's cold warriors spent in front of the screens.

While the American military worked on SAGE, computers were also being developed by the main adversary of the United States. The Soviet Union built its first electronic digital computer in 1950 at the Kiev Institute

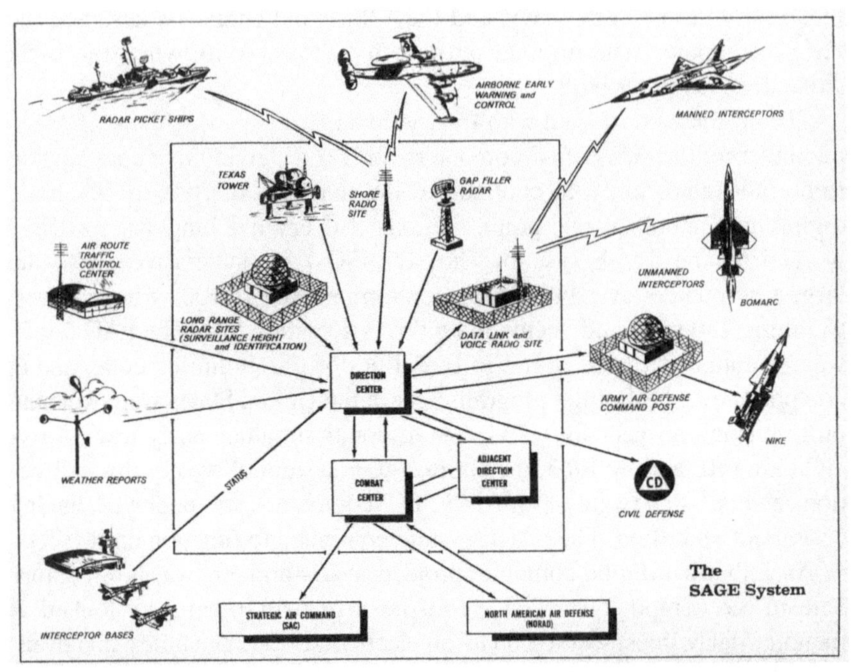


Design of the complex SAGE system. (Image Courtesy of the Charles Babbage Institute Archives, University of Minnesota Libraries, Minneapolis)

of Electric Engineering under the direction of S. A. Lebedev (1902–1974), which ran fifty instructions a second and used a memory of thirty-one 16-bit words. After moving to Moscow, Lebedev proceeded to supervise the development of the BESM (which stood for "large electronic computer" in Russian) series of computers that used magnetic drums and magnetic tapes for storage. By and large, Soviet computers were copies of American computers, purchased illegally through third-party countries, since during most of the Cold War the United States forbade the export of computers to Soviet companies, considering computers to be as lethal as arms or munitions.

## TRANSISTORS

Early computer designers were tormented by the nature of vacuum tubes, constantly struggling to deal with excess heat, their bulky size, and their penchant for failure. The room-sized dimensions of computers came directly from efforts to cope with vacuum tubes. Salvation came from Bell

Telephone Laboratories in Murray Hill, New Jersey, where the theoretical physicist John Bardeen (1908–1991) and experimental physicist Walter H. Brattain (1902–1987) invented the point-contact transistor, made out of germanium, a semiconductor material, in 1947. The physicist William B. Shockley (1910–1989) was the team leader for the project but was not involved in the initial invention. Shockley took the invention and made the junction transistor, which became the basis for later commercial development. Transistors acted as a switch to turn the flow of electricity on or off and as an amplifier to the current. The American military immediately recognized the long-term value of transistors and provided additional funding to further develop the invention.

Though the three Bell scientists shared the 1956 Nobel Prize in Physics for their invention, rivalries among the three drove them apart. In 1956, Shockley returned to Palo Alto, California, where he had grown up, and founded Shockley Transistor. Under the influence of nearby Stanford University, the

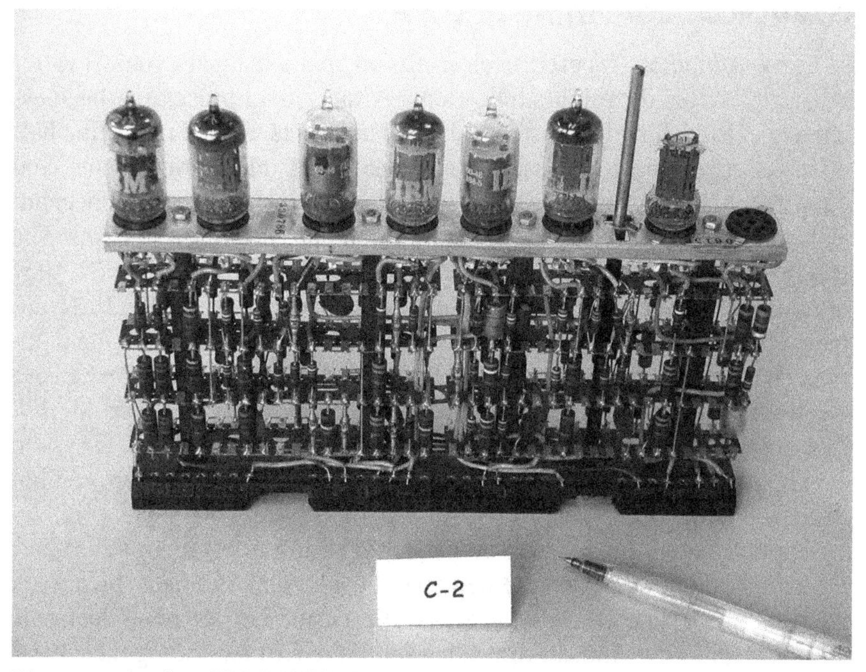


Photograph of an IBM 1955 IBM vacuum tube logic unit (or gate) from an IBM 705. (Image Courtesy of the Digital History Archive at Weber State University)

surrounding towns already made up a nascent center of technological innovation. Shockley's decision to locate there eventually led to the area becoming known as Silicon Valley. Shockley intended to create commercial transistor products, but his abusive management style alienated his employees. As we explore below, eight of his employees fled to form their own company, Fairchild Semiconductor, a year later. After Shockley's company faded away, he turned to the study of genetics and became infamous for promoting racist views of human inheritance.

By the late 1950s, the transistor had become a useful commercial product, creating the second generation of computer hardware and rapidly replacing vacuum tubes in computers and other electronic devices because transistors were much smaller, generated less heat, and were more reliable. Defense-related projects and space-related projects in the United States, as part of the Cold War, became a major driver for computer-related innovation in both hardware and software.

## COMMERCIAL COMPUTING AND IBM

Early computers focused exclusively on mathematical processing for scientific and engineering applications. A large data processing business already existed, having transformed how businesses worked in the first half of the twentieth century. Insurance companies, banks, governments, and other organizations that processed large amounts of data relied on mechanical calculators, punched cards, tabulators, and physical filing systems. The UNIVAC electronic computers promised to revolutionize this market. IBM dominated the market for punched card equipment, and in 1951, the company changed its sales predictions to include a substantial emerging commercial market for electronic computers. IBM already had extensive electronic computing experience through the work on the Harvard Mark I. The contract to produce the AN/FSQ-7 computers for the SAGE system also helped IBM catch up on the technology and even to develop leading-edge technologies.

Approximately ten large computer companies emerged in the 1950s, all competing to dominate the new market. The United States, flush with postwar prosperity, provided most of the customers for new electronic computers, as other industrialized nations were still rebuilding their basic infrastructure after the devastation of World War II. The new computer companies were usually preexisting electronics manufacturers or business machine manufacturers expanding into a new area and included General

Electric, RCA, Raytheon, Honeywell, Burroughs, Remington Rand, Monroe, and Philco. IBM had an advantage over other companies because of its highly motivated sales force that was accustomed to developing complete data processing solutions for its customers, not just selling or leasing equipment.

In the early 1950s, as commercial companies began to build production computers, following the example of the UNIVAC, there were many one-of-a-kind computers built, such as the IAS by John von Neumann at the Institute for Advanced Study at Princeton University. The National Applied Mathematics Laboratories, part of the National Bureau of Standards, built the Standards Eastern Automatic Computer (SEAC) in its Washington, DC, office, and the Standards Western Automatic Computer (SWAC) in its Los Angeles office, both in 1950. The National Physical Laboratory in Teddington, Middlesex, England, built its Pilot Automatic Computing Engine (ACE) in 1950, which was designed by Alan Turing. IBM delivered the Naval Ordnance Research Calculator (NORC) for the U.S. Navy's Bureau of Ordnance in 1954. Though the NORC was built for its actual cost, with the company not charging extra for profits, IBM benefited from technical advances that it applied to later computers. All these computers were based on vacuum tube technology and were designed for scientific calculations, not data processing.

While working on both the NORC and AN/FSQ-7 computers, IBM also canvased the market for further government, defense-related, and aviation-related customers. It found thirty customers willing to sign letters of intent and began work on what became known as the IBM 701. The 701 drew heavily on the design of the Princeton IAS, and used Williams tube electrostatic memory units developed at Manchester University to achieve a total capacity of 4,096 36-bit words. The original memory units were visible through glass doors, allowing computer operators to see the contents of each memory location as dots on the cathode-ray tubes (CRTs). A photographer at the official unveiling of the computer in 1952 used a flashbulb, which promptly reset the tube-based memory with random bits. The IBM engineers later built memory units with darkened glass covers.

IBM brought its strategy of leasing equipment, rather than selling it, from its punched card business into its electronic computer business. The 701 leased for $8,100 a month. Because the 701 proved more expensive than initially projected, only nineteen customers leased the machines. Douglas Aircraft Company purchased two to help design such noted aircraft as the DC-6B, DC-7, and DC-8 aircraft as well as the A-3D Skywarrior and A-4D Skyhawk aircraft for the U.S. Navy. Lockheed Aircraft Company also purchased two IBM 701 computers. The users of the IBM 701 formed a group called SHARE, which became an informal mechanism for exchanging programs among themselves and pressuring IBM to create more software products. Everyone already recognized that one of the advantages of common machines was that programs did not have to be rewritten for each individual computer. User groups for computers from other manufacturers also became common. These user groups provided a way for customers to unite and encourage manufacturers to implement preferred features and develop particular types of software. Chapter meetings and annual conferences became a way for the manufacturer to communicate with customers and for people with job skills on a particular computer model to find a job with another company with that same computer model.

IBM followed up the 701 with its 704 computer. Brought to market in 1954, the 704 was the first mass-produced computer to use magnetic-core memory and offer built-in floating-point hardware for handling decimal numbers. IBM also improved the 704 over the 701 by providing a set of seventy-five basic instructions for the 704, whereas the 701 only had thirty-three instructions. The greater number of instructions made it easier for programmers to do more complex operations. The 704 also used a rotating magnetic drum for secondary memory.

The IBM 702, which completed development a year after the 704, was not designed as a fast scientific calculator, like earlier computers, but for business data processing. To facilitate data processing, which emphasized transactions of characters and numbers, the memory of the 702 was oriented toward storing characters. The 702 also included a new tape drive that successfully used plastic-based magnetic tapes and prompted the entire industry to switch away from the metal-based tapes that the UNIVAC had pioneered. After producing only 14 machines, the 702 was superseded by the 705, which IBM continued to upgrade until it reached 80,000 characters of memory in 1959, and the company leased 180 of the machines.

IBM also introduced the 650 Magnetic Drum Data Processing Machine in 1954. It was designed as a business data processing machine midway between the punched card systems still in wide use and the processing power of the IBM 702. The 650 used a rotating drum to store 2,000 ten-digit words, which contained both instructions and data. Though the drum spun at 12,500 revolutions per second, the drum took about 5 milliseconds to complete a rotation. Many computer instructions took only about 3 milliseconds to execute, so when instructions were laid sequentially across the drum, about 2 milliseconds were wasted during every execution cycle.

The solution was to carefully stagger the instructions of a program across the drum to minimize the latency between instructions. The 650 leased for $3,250 a month, and IBM eventually sold 2,000 of these machines, earning more money than from its entire 700 series. With a canny understanding of market dynamics, IBM offered deep discounts on 650 computers to institutions of higher education, as long as the college or university began to teach programming. Of course, the students learned to program IBM computers. The IBM 650 established IBM's lead in the electronic computer industry. In 1955, the orders for just the more flashy 700 series computers exceeded the orders for all Sperry Rand UNIVACs.

The economics of the time meant that computers were expensive, and anything that made their use more efficient was desirable. A major problem for efficiency came from the fact that peripheral devices were so much slower that the main central processing unit (CPU), and the CPU was often idle while a tape was being read or written or while data was being sent to a line printer. IBM introduced its smaller 1401 computer, whose sole task was writing data to a magnetic tape or copying data from a tape and sending it to a line printer, and eventually sold over 12,000 IBM 1401 model computers. IBM complemented the 1401 by developing the 1403 printer, which produced 600 lines per minute. The printer used a horizontal chain with characters on it that moved rapidly back and forth while hammers slammed the characters on the chain into an ink ribbon, leaving a printed character on the paper. IBM eventually sold more than 23,000 of these printers, keeping them available for sale until 1983.

General Electric (GE) got its start in the computer industry by building specialized computers for the military. In the mid-1950s, W. R. G. "Doc" Baker (1892–1960), an executive at GE, became aware of a commercial opportunity at Bank of America (BoA). Concerned about its ability to efficiently handle the millions of checks that flowed through its banks every year, BoA contracted with the Stanford Research Institute (SRI) to build a prototype machine that processed checks and recorded the necessary accounting information. SRI developed magnetic ink character recognition (MICR) so that magnetic characters imprinted on the checks could be read by the electromechanical check readers. Burroughs also developed a technology using fluorescent dots and IBM preferred to use barcodes to solve the problem of how to get the check reader to quickly read the checks. The SRI machine, finished in 1955, included 8,000 vacuum tubes, 34,000 diodes, and a million feet of copper wire. The machine was hardwired and did not use any form of software or programming.

GE won the bid on the contract to build thirty-six production models of this machine at a cost of $31 million. The natural choice to win the bid, IBM, did not bid, but tried instead to buy the idea from BoA and SRI so that IBM could develop it into a general solution to sell to other banks. BoA chose not to sell, and GE went to work. The proposed Electronic Recording Machine Accounting (ERMA) computer was completely redesigned and used 5,000 transistors, 15,000 diodes, and 4,000 resisters, turning it from a hardwired computer into a stored-program computer. An early programming language, Intercom 100, was developed just for these computers. In late 1958, the first computer was placed in a bank and was only able to process 100 banking transactions per day. After further refinement, in just three months, the computer achieved the 55,000 transactions per day required by the contract. The check sorter and check reader, developed by the National Cash Register Company (NCR), could even handle checks that had been crumpled and stepped on before being smoothed out by hand and set in the sorter. GE renamed ERMA the GE 100 and entered the commercial computer industry in earnest, though the company's leadership always remained leery of competing with IBM head to head.

As always, American defense spending remained an important driver in the first three decades of electronic computers. In 1955, the Los Alamos Scientific Laboratory of the U.S. Atomic Energy Commission requested a computer 100 times faster than any then in existence, and IBM decided to take up the challenge. Instead of building a one-of-a-kind machine like the previous NORC, IBM designed the computer to meet Los Alamos' requirement and to become a new commercial computer, the IBM 7030. The project was appropriately named Stretch. The new computer used transistors, faster magnetic-core memory, and the pipelining of instructions. Pipelining is when electronic circuitry in the CPU is designed to not only execute the current instruction but also to begin decoding and processing subsequent instructions at the same time. The 7030 decoded and partially processed the five instructions beyond the current instruction and was ready to throw away the extra work if an earlier instruction proved to be a branch or to make some change that made the work of the later instructions invalid. The 7030 also had a memory controller that prefetched data from memory locations before the CPU asked for them, anticipating the needs of the program. IBM pioneered random-access disk drives, and the drive for the 7030 was the first drive to include more than a single read/write arm in the same movable system, leading to much greater storage capacity in hard drives.

The 7030 was the most complex computer ever built up to that time, and IBM engineers used programs on an IBM 704 to simulate the features of the system, especially the pipelining features. Computers were now being used to design new computers. Though the Stretch project did not quite meet its goal of a hundredfold increase in performance, the first 7030 was delivered to Los Alamos in 1961, and seven others were produced for customers in England, France, and the United States. One of the machines became the core of the Harvest machine for the U.S. National Security Agency (NSA), the government agency tasked with codebreaking.

The pace of technological innovation forced IBM to come up with ever newer machines. The 709 was first produced in 1958 as a successor to the 701 and 704 series. IBM took care to make the 709 downward (or backward) compatible, which meant that programs written for the early machines could run without modification on the newer machine. Philco Corporation introduced the Transac S-2000 in 1958, one of the first commercial computers built with transistors instead of vacuum tubes. A new model of the UNIVAC was also being prepared using transistors. IBM quickly reacted by building a new version of the 709, the 709TX, using transistor technology. The 709TX was designed for use in the Ballistic Missile Early Warning System (BMEWS), a complement to the SAGE system. Whereas SAGE detected the previous threat from aircraft, the BMEWS sought to detect the new threat of nuclear-tipped intercontinental ballistic missiles (ICBMs). IBM converted the 709TX into its 7090 computer for commercial customers, which ran five times faster than the 709 and had much greater reliability because of the use of transistors instead of vacuum tubes.

By the end of the 1950s, IBM had a wide range of computers to meet the different needs of the scientific and data processing industries. This large number of systems confused the marketplace and fragmented the technical efforts at IBM, yet IBM had come to dominate the industry to such an extent that people took to calling the situation "Snow White and the Seven Dwarfs," where IBM was Snow White and a slew of smaller companies, not always seven in number, tried to survive in IBM's shadow. IBM's industrial designers also created a light blue design for the computer cabinets, earning the company the nickname "Big Blue." The company employees also took to wearing blue suits and white shirts.

Despite its dominance of the electronic computer industry, in 1959, the majority of IBM's revenue still came from its punched card business, not computers. Overseas sales for IBM were also important, with 20 percent of company revenue in 1960 coming from outside the United States. This increased to 35 percent in 1969, 54 percent in 1979, and 61 percent in 1990. The American computer industry provided computer hardware and software for the rest of the world, usually pushing local companies to the margins.

Only Sperry Rand and its UNIVAC computers were successful enough to continue to directly compete with IBM. Whereas IBM had built the 7030 in its Stretch project, Sperry Rand built the Livermore Atomic Research Computer (LARC) for the Lawrence Radiation Laboratory of the Atomic Energy Commission in California. The LARC was comparable to the 7030, though it used high-speed drums rather than the newer disk drive technology. After delivering the LARC in 1960, Sperry Rand built a second one for the U.S. Navy Research and Development Center in Washington, DC.

The Stretch project and the LARC were supercomputers compared to contemporary machines. The third supercomputer project came from Ferranti in England, where Tom Kilburn directed a team to build the Ferranti Atlas. The Atlas pioneered two important technologies: virtual memory and some aspects of time-sharing. The Atlas was designed to use a memory space of up to a million words, with each word being 48 bits long. No one could afford to put that much magnetic-core memory in a machine, so the Atlas had actual core memory of only 16,000 words. A drum provided 96,000 more words. The operating system of the Atlas swapped content from its magnetic-core memory to and from the more cost-effective drum in the form of pages, providing the illusion of more memory via this virtual memory scheme. The Atlas was also designed to be a time-sharing computer so that more than one program could be run at a time. To implement this time-sharing, the idea of extra code was developed, which is similar to what is now called *system interrupts*. These two ideas were adopted in all later operating systems of any sophistication. As with the other supercomputers, the Atlas was not a commercial success, as only three were built.

## PROGRAMMING LANGUAGES

The first programs were created by rearranging the plugs and wiring on the computer. The von Neumann idea of using memory to hold the program instructions and data, not just the data, required that another method of entering the program be invented. Punched cards, paper tape, and magnetic tape were soon adopted. The first programs were written in straight binary, also called *machine code*, and were difficult to write and debug. What programmers of the time called *automatic programming* soon emerged, mainly in an effort to reuse binary code that performed common functions, such as floating-point arithmetic, which is calculations using decimal numbers. As these automatic programming systems grew more complex, they frustrated programmers because the resulting code was not as efficient and ran five to ten times slower than programs written from scratch.

The economics of computing costs had already emerged in the early 1950s; programmer and computer operator salaries had become the major costs of a computer center. A common feature of high technology is that people cost more than the machines that they use. Managers found that when computers were used for both development and production activities, most of the computer time was taken up by programming and debugging activities rather than getting production work done. Anything that increased the efficiency of programmers helped the productivity of the computer center. But what if the programs created via automatic programming ran so slowly that the economic advantage in their method of creation was lost during the years that the programs ran?

Programmers attacked the problem by creating higher-level computer languages. Some early examples of these languages were algebraic compilers, with names such as Short Code, Mathe-Matic, Speedcoding, and Autocode. The IBM 704 had floating-point logic built into its circuits so that programmers no longer had to write code to manipulate decimal numbers. John Backus (1924–), an IBM employee, saw an opportunity. In 1954, Backus proposed to create a new programming language that made it easier to write computer programs but was efficient enough to compete with hand-coded binary programs. IBM gave Backus a small team of programmers, and they set out to create FORTRAN (an acronym for FORmula TRANslation). The team completely focused on creating efficient object code and literally made up the language as they went along, in contrast to most later programming languages, which were planned out before development. The team also decided to ignore any blanks in the source code, since blanks always caused a lot of problems for people running keypunches.

The FORTRAN project took longer than anticipated and was not completed until April 1957. The initial release of the compiler came in a deck of 2,000 punch cards. The language became popular with users of the IBM 704, and the language rapidly went through new versions that added new features, especially features that helped with debugging. FORTRAN II came out in the spring of 1958, followed by FORTRAN III and FORTRAN IV, with the FORTRAN 66 definition coming out in 1966. The language was also ported to other computer systems besides the IBM family of computers and served as the basic programming language for scientific and engineering applications for decades. FORTRAN was simple enough that scientists and engineers could learn to write programs themselves.

While FORTRAN served the scientific and engineering segment of computer users, business users were becoming ever more important. Business users often wrote programs that processed data in the form of transactions, for instance, reservations for airlines or checks being processed and cleared. The U.S. Department of Defense also had a need for data processing programs and sponsored the creation of COBOL (for Common Business-Oriented Language). Grace Hopper helped lead the effort to create this language and remained heavily involved with navy computer programming efforts. The navy waived its mandatory retirement age for her, and she eventually retired in 1986, after attaining the rank of rear admiral. Some programmers considered COBOL too wordy, but other programmers lauded the readability of using complete English words, like MULTIPLY and READ, rather than the terse syntax of FORTRAN.

FORTRAN and COBOL are considered third-generation computer languages. They followed the assembly languages that make up the second generation and the binary machine code of the first generation. The historical generations of programming languages do not correspond with the historical generations of computing hardware. For instance, COBOL and FORTRAN were written on second-generation hardware based on transistors. IBM also developed Report Program Generator (RPG), a simple language used to quickly generate simple business and accounting applications. The language was designed to mimic punched card machines so that people already trained in using plugboards on punched card machines could easily transfer their arcane expertise to the computer. Plugboards that looked like switchboards trace back to the ENIAC and were a mechanical means of programming the machine. Programmers would actually store the plugboards with all their properly placed wires and simply plug the prewired plugboards into the machine to run a specific program. Punch cards alleviated using that clumsy hardware. Other languages also emerged at the time, such as List Processing (LISP) in 1958 and Algorithmic Language (ALGOL) in 1960.

## SOFTWARE SYSTEMS

As the computer industry developed during the 1950s, IBM began to develop suites of software aimed at particular industries, such as banks, manufacturing, and insurance. IBM gave away the software to those industries as an incentive to buy computers, computer peripherals, and services. Other computer companies did not have the deep financial resources of IBM and concentrated on one or two classes of customers, developing software to give away so that they could compete with IBM in that more restricted arena. Burroughs and GE concentrated on banking, NCR concentrated on retailing, and Control Data Corporation concentrated on scientific computing.

In 1955, two programmers at IBM left to form the first computer software services company, the Computer Usage Company (CUC). The company's first project simulated the radial flow of fluids in an oil well for the California Research Corporation. Other contract programming projects followed. The RAND Corporation, a think tank owned by the U.S. government, created a subsidiary, Systems Development Corporation (SDS), in 1957 to write computer programs for the SAGE air defense project. Computer Sciences Corporation (CSC) was formed in 1959, and its first contract was to develop a business-language compiler for Honeywell. CSC initially worked on writing systems software for computer manufacturers but later focused on providing computer contracting services to the federal government and military. Software contracting firms also emerged in Europe, usually a few years later than American examples. For example, in 1958, Banque de Paris and Marcel Loichot formed the Sema company in France.

Until the 1960s, all software—with the exception of some system software, such as operating systems, interpreters, compilers, and system utilities—was developed for individual applications. All of the early computer software companies relied on contract programming for their income because a software market had not yet developed to sell their software independently. Many of these early software systems automated previous manual business processes or business processes that had been partially automated by punched card systems.

An example of an early manual data processing system, the Reservisor system at American Airlines, was used to arrange airline reservations. As the jet age began, causing explosive growth in the airline industry, American Airlines planned to expand the Reservisor system but realized that its manual methods would no longer function on the scale required. IBM became involved in finding a solution for American Airlines. Realizing how long a computerized system would take to develop, IBM created a temporary system based on punched card processing for use in the late 1950s. The computerized SABRE system, which was begun in 1957, was finally completed in 1964 and was based on two IBM 7090 mainframes running a million lines of code. With the largest online storage system up to that time, 800 megabytes in size, SABRE supported 1,100 travel agents around the country connected to the system via terminals. With updated hardware, SABRE has continued run to this day, standing as an example of a large real-time transaction-processing system with sophisticated measures to avoid any downtime. SABRE became the basis for the well-known Travelocity website (http://www.travelocity.com), which was created in 1996 with SABRE providing the background data processing engine and the new web technology providing a way to directly reach customers with an easy-to-use interface.

## EARLY EFFORTS AT ARTIFICIAL INTELLIGENCE

Early computer pioneers, like the mathematicians Alan Turing and John von Neumann, intentionally developed electronic computers as the first step toward the creation of genuine thinking machines. In 1950, Turing created the *Turing test*, or *imitation game*, which proposed how to test an intelligent machine. The test required a human to have a conversation with a computer via a teletype terminal, and if the human could not determine whether the answers came from a human or a computer, then artificial intelligence had been achieved. Of course, this test would have to be conducted many times with many people before that final conclusion was reached. To date, no computer or program has come close to passing the Turing test. While contemporary researchers no longer commonly accept the Turing test as an absolute criterion, it remains a good indicator of the goals of these early pioneers.

The mathematician Marvin Minsky (1927–2016) became one of the pioneers of research in artificial intelligence. Minsky was born in New York, where his father was an eye surgeon and his mother was a Jewish activist. Minsky attended private schools, where his gifts were recognized, and after brief service in the U.S. Navy, he entered Harvard University in 1946. Initially majoring in physics, he expanded his interests into psychology after becoming fascinated with how the mind worked and graduated in 1950 with a BA in mathematics. Minsky then attended Princeton

University, and in 1951, he and a colleague built the Stochastic Neural Analog Reinforcement Computer (SNARC). Made out of 400 vacuum tubes, the machine was an early attempt at creating a learning system based on neural nets like those of the human brain. Minsky earned a PhD in mathematics from Princeton in 1954 and returned to Harvard as a junior fellow, where he worked on microscopes and patented a scanning microscope.

Minsky and John McCarthy (1927–2011) organized a two-month summer workshop at Dartmouth College in 1956, where the term *artificial intelligence* (AI) was first coined. Artificial intelligence research became defined as the effort to create computer hardware and software that behaved as humans do and that actually thought. In 1958, Minsky and McCarthy joined the faculty of the Massachusetts Institute of Technology (MIT), and a year later, they founded the MIT Artificial Intelligence Project, which became the Artificial Intelligence Laboratory in 1964. In 1956, upon learning that IBM planned to donate one of its 704 computers to MIT and Dartmouth, McCarthy conceptually developed a new programming language for use in AI research before the computer even arrived. His algebraic list-processing language became LISP in 1958 and is still used in AI research.

The first two decades of AI research were dominated by researchers at MIT, Carnegie Tech (later renamed Carnegie Mellon University), Stanford University, and IBM. However, European and Japanese efforts later became important. The history of AI has been characterized by a series of theories that showed initial promise when applied to limited cases—leading to optimistic declarations that intelligent machines were just around the corner—but that brought disappointment as the theories failed when applied to more difficult problems, as we will see in greater detail later.

# FOUR

# The Third Generation: From Integrated Circuits to Microprocessors

## INTEGRATED CIRCUITS

Jack S. Kilby (1923–2005) grew up in Great Bend, Kansas, where he learned about electricity and ham radios from his father, who ran a small electric company. Kilby desperately wanted to go to the Massachusetts Institute of Technology (MIT), but he failed to qualify when he scored 497 instead of the required 500 on an entrance exam. He turned to the University of Illinois, where he worked on a bachelor's degree in electrical engineering. Two years of repairing radios in the army during World War II interrupted his education before he graduated in 1947. He moved to Milwaukee, Wisconsin, where he went to work for Centralab. A master's degree in electrical engineering followed in 1950. Centralab adopted transistors early, and Kilby became an expert on the technology, though he felt that germanium was the wrong choice for materials. He preferred silicon, which could withstand higher temperatures, though it was more difficult to work with than germanium. After Centralab refused to move to silicon, Kilby moved to Dallas to work for Texas Instruments (TI) in 1958.

TI had been founded in the 1930s as an oil exploration company and later turned to electronics. It produced the first commercial silicon transistor in 1954 and designed the first commercial transistor radio that same year. In his first summer at the company, Kilby had no vacation days accrued when everyone else went on vacation, so he had a couple of weeks of solitude at work. TI wanted him to work on a U.S. Army project to build micromodules in an effort to make larger standardized electronics modules. Kilby thought the idea ill conceived and realized that he had only a short time to come up with a better idea.

Transistors had transformed the construction of computers, but as ever more transistors and other electronic components were crammed into smaller spaces, the problem of connecting them together with wires required a magnifying glass and steady hands. The limits of manufacturing electronics by hand became apparent. Making the electronic circuitry larger by spacing the components farther apart only slowed the machine down because the electrons took more time to flow through the longer wires.

Transistors were built of semiconductor material, and Kilby realized that other electronic components used to create a complete electric circuit, such as resistors, capacitors, and diodes, could also be built of semiconductors. What if he could put all the components on the same piece of semiconductor material? On July 24, 1958, Kilby sketched out his ideas in his engineering notebook. The idea of putting everything on a single chip of semiconductor later became known as the *monolithic* idea, or the *integrated circuit.*

By September, Kilby had built a working integrated circuit. To speed up the development process, Kilby worked with germanium, though he later switched to silicon. His first effort looked crude, with protruding gold wiring that Kilby used to connect each component by hand, but the company recognized an important invention. Kilby and TI filed for a patent in February 1959.

In California, at Fairchild Semiconductor, Robert Noyce (1927–1990) had independently come up with the same monolithic idea. Noyce graduated with a doctorate in physics from MIT in 1953 and then turned to pursuing his intense interest in transistors. Noyce worked at Shockley Transistor for only a year, enduring the paranoid atmosphere as the company's founder, William B. Shockley (who had been part of the team that invented the transistor in 1947), searched among his employees for illusionary conspiracies. Noyce and seven other engineers and scientists at the company talked to the venture capitalist who provided the funding for the company but could obtain no action against Shockley, a recent Nobel laureate. Shockley called the men the "traitorous eight" when Noyce and his fellow rebels found financing from Fairchild Camera and Instrument to create Fairchild Semiconductor.

Fairchild Semiconductor began to manufacture transistors and created a chemical planar process that involved applying a layer of silicon oxide on top of electronic components to protect them from dust and other contaminants. This invention led Noyce to develop his own idea of the integrated circuit in January 1959, using the planar process to create tiny lines of metal between electronic components to act as connections in a semiconductor substrate. After TI and Kilby filed for their patent, Noyce and Fairchild Semiconductor filed for their own patent in July 1959, five months later. The latter patent application included applying the chemical planar process.

Kilby and Noyce always remained friendly about their joint invention, while their respective companies engaged in a court fight over patent rights. Eventually the two companies and two engineers agreed to share the rights and royalties, although the U.S. Court of Customs and Patent Appeals ruled in favor of Fairchild in 1969. The two companies agreed, as did the two men, that they were coinventors. Kilby later received half of the 2000 Nobel Prize in Physics for his invention, a recognition of its importance. Noyce had already died, and Nobel Prizes are not awarded posthumously. The other half of the Nobel Prize was shared by the Russian physicist Zhores I. Alferov (1930–2019) and German-born American physicist Herbert Kroemer (1928–) for their own contributions to semiconductor theory.

## NASA

It is hard to overestimate how much America's space program drove computer development generally in network connectivity, software, and hardware miniaturization. It did this while still needing to utilize existing and established practices at an enormous scale to create resilient systems. The commercial electronics industry did not initially appreciate the value of integrated circuits (also called *microchips*), believing them too unreliable and too difficult to manufacture. However, both the National Aeronautics and Space Administration (NASA) and the American defense industry recognized the value of microchips and became significant early adopters, proving that the technology was ready for commercial use. The U.S. Air Force used 2,500 TI microchips in 1962 in the onboard guidance computer for each nuclear-tipped Minuteman intercontinental ballistic missile. NASA used microchips from Fairchild in its Project Gemini in the early 1960s, a successful effort to launch two-man capsules and to rendezvous between a manned capsule and an empty booster in orbit. NASA's $25 billion Apollo program (or Project Apollo) to land a man on the moon became the first big customer of integrated circuits and proved a key driver in accelerating the growth of the semiconductor industry. At one point in the early 1960s, the Apollo program consumed over half of all integrated circuits being manufactured.

Networking took a giant leap forward because NASA needed to create a system that tracked and communicated with the spacecraft in terms of data, voice, and television while it orbited the earth and headed for the moon, a quarter of a million miles away. This required fourteen linked antennas located around the globe, from Florida to Australia, as well as four ships at sea and eight aircraft in the air. They were linked by undersea cables and microwave towers as well as using the telephone system. The bills for using the telephone system alone were tens of millions of dollars. The Spaceflight Tracking and Data Network (STDN) took 2,700 people to operate and operated at speeds of 52K bits per second (bps) from the spacecraft but a mere 2K bps in the other direction. That slow uplink speed is one reason why, unlike much of the rest of humanity, Michael Collins, who orbited the moon while Neil Armstrong and Buzz Aldrin took man's first steps on the moon during Apollo 11, could not see those images. The speed between Goddard Space Flight Center in Maryland and Mission Control in Houston was 41K bps, a speed typical consumers did not see through their modems for another three decades.

Apollo also led to the development of the idea of software engineering, which is applying sound engineering principles of verification to the development of computer programs to ensure that they are reliable in all circumstances. The prior dramatic explosive loss of the unmanned Mariner 1 made that clear. The U.S. government designed the Mariner spacecraft series to be the first interplanetary spacecraft, carrying cameras and scientific instruments, powered by solar panels, and radioing their findings back to Earth. An Atlas-Agena B rocket launched the first Mariner spacecraft on July 22, 1962. A guidance signal from the ground directed the rocket's ascent. When that signal failed, an onboard computer took over. A programming mistake in the FORTRAN program running on the onboard computer left out a character from the computer program, causing the rocket to veer to the left. A range safety officer detonated an explosive charge within the rocket to prevent any harm to people watching the launch. Later Mariner spacecraft, with the bug fixed, successfully visited the planets Venus, Mars, and Mercury. The missing character (called a "hypen" by the press) would have acted as an averaging function for a variable and would have allowed the software to allow for momentary lapses of communication with the ground. Science fiction author Arthur C. Clarke called it the "most expensive hyphen in history."

The Mariner experience changed the way NASA developed and tested software. Margaret Hamilton (1936–), at MIT's Instrumentation Laboratory, has regularly been credited with either inventing or popularizing the phrase *software engineering* to distinguish it from an "art." In 1966, many at MIT thought the phrase was a joke. By 1969, she was in charge of all software for the Apollo command module. The software ran on a computer in real time with limited memory. It had over 200 inputs and managed all the inputs by being one of the first systems in the world that had decision-making capabilities, prioritizing the most important decisions. The program could also fail and recover gracefully because it stored what it was working on and reset to that level if it failed. This resilience was put to the test when the Apollo 11 lunar lander's computer rebooted five times in four minutes on its approach to the moon. It flashed a 1201 error because a radar unit had accidentally been turned on, and the information from that unit had flooded the computer's tiny memory. NASA ground controllers did not know why the error occurred until later, but they did not abort the mission because controllers noticed that the system kept working despite the error message. That resilience led some at NASA to call the computer the "fourth crew member."

The MIT Instrumentation Laboratory, working with the Raytheon Corporation, played an important role in miniaturization in building the computer that ran the critical navigation software. The lab was run by Charles Stark Draper (1901–1987), who had created the navigation systems for the U.S. Navy's nuclear submarines. The navigation systems aboard both the command and lunar modules needed to be the size of a briefcase—far smaller than the refrigerator- to room-sized computers at the time. The computers needed to run in real time, and even though the computers were small, the controls needed to be handled by astronauts wearing bulky gloves. MIT managed this by gambling on a new technology, integrated circuits, and this led to pushing that new hardware.

NASA and military requirements also led to advances in the creation of fault-tolerant electronics. If a business computer failed, a technician could repair the problem, and the nightly accounting program could continue to run; if the computer on the Saturn V rocket failed, the astronauts aboard might die. Electronics were also hardened to survive the shaking from a rocket launch and exposure to the harsh radiation and temperatures of space. It became obvious that the weight and size of transistors would not work. Unfortunately, in 1962, an integrated circuit holding a mere six

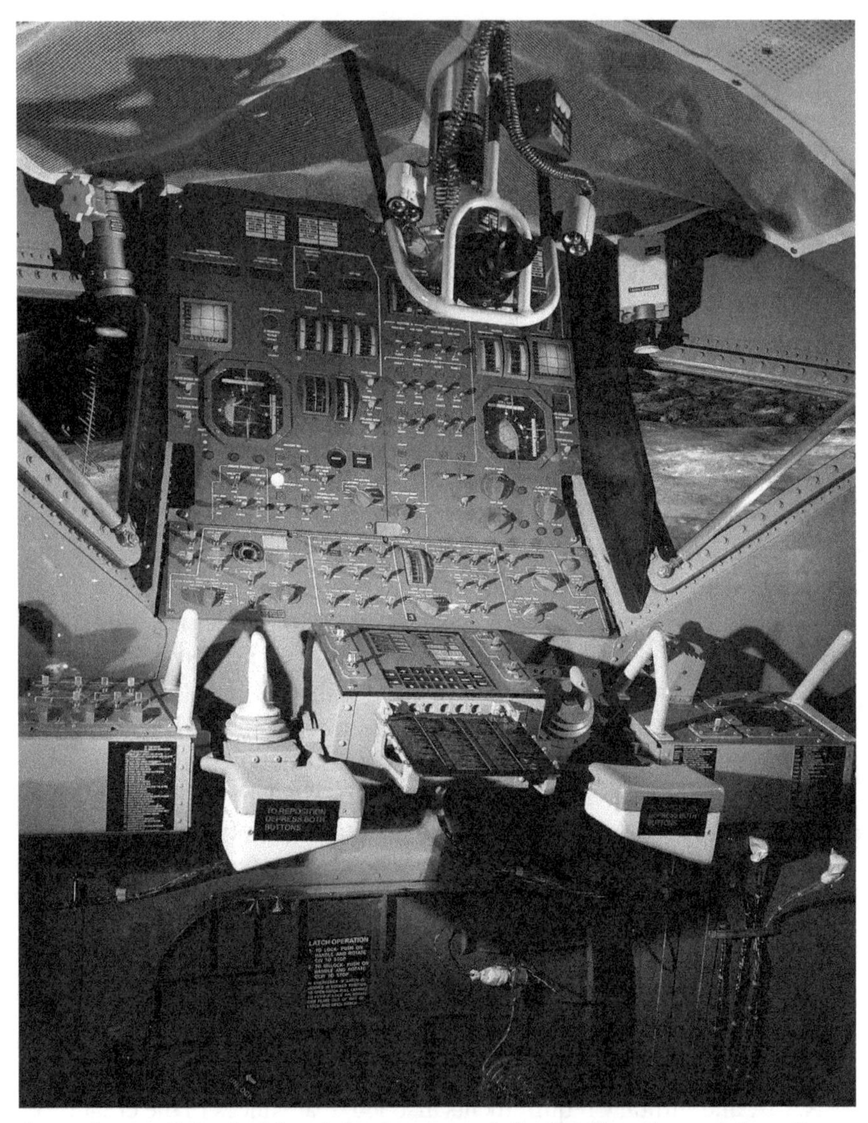


A mock-up of the inside of the lunar module (LM) and control surfaces. The interface to the computer is in the center of the image. The computer helped spur the development of miniaturization and software engineering. (Philcold/Dreamstime.com)

transistors cost $1,000 and frequently failed. By buying in bulk, using a battery of tests, and rejecting whole lots, NASA drove the price down to $15 apiece within a year while driving up capability and reliability. It was NASA's purchase and the subsequent development of microchips that inspired Gordon E. Moore (1929–) to write Moore's Law, which we explore below, in 1965. Meanwhile, the mid-1960s development of large computers like the IBM 360 continued with transistors. The use of microchips was a huge gamble for NASA but one that paid off for them and the entire computer industry. By the end of the 1960s, IBM and other companies could take advantage of the integrated circuit.

While software and hardware advanced, the memory requirements for the Apollo mission settled on an older technology—core rope memory assembled by hand by women recruited from the textile factories of Waltham, Massachusetts. The flight software was loaded onboard months before the flight. The technique did have the advantage of being resilient and unlikely to be erased by errant work by technicians, power surges, or acts of nature. The Apollo 12 mission survived two lightning strikes on its way to orbit without damaging the onboard computer system. Core memory still had the densest memory per weight capacity of any technique of the day when introduced into the Apollo system. But on land, microchips and magnetic media were supplanting core memory by the time Apollo 11 headed for the moon in 1969.

## MOORE'S LAW

Commercial industries began to appreciate the value of integrated circuits when Kilby and two colleagues created the first electronic calculator using microchips in 1967. The calculator printed out its result on a small thermal printer that Kilby also invented. This was the first electronic calculator small enough to be held by hand and sparked what became a billion-dollar market for cheap handheld calculators, quickly banishing slide rules to museums. Integrated circuits became the main technology of the computer industry after a decade of development, creating a third generation of computer technology (following the first and second generations based on vacuum tubes and transistors, respectively).

In 1964, Gordon E. Moore noticed that the density of transistors and other components on integrated chips doubled every year. He charted this trend and predicted that it would continue until about 1980, when the density of integrated circuits would decline to doubling every two years. Variations of this idea became known as Moore's Law. Since the early 1970s, chip density on integrated circuits, both microprocessors and memory chips, has doubled about every eighteen months, from about fifty electronic components per microchip in 1965 to forty-two million electronic components per microchip in 2000. Moore also pointed out another way to understand the growth of manufactured semiconductor material. From the beginning, every acre of silicon wafer has sold for about a billion dollars; the number of transistors and other electronic components on the chip have just become denser to keep the value of that acre roughly constant. The amount of attention paid to density paid dividends over time. While the manufacture of integrated circuits is considered to be part of the electronics industry, the industrial techniques used are more like those of the chemical industry. A mask is used to etch patterns into wafers of silicon, creating a large number of integrated circuits in each batch. The key to economic success came from high yields of mistake-free batches.

Central processing unit (CPU) microchips had so many extra transistors that designers began to build multiple CPU cores on a single chip, each core an individual CPU, allowing programs to run in parallel. Many types of programs were not capable of running parallel processes or multiple threads because each step of the program required the completion of earlier steps of the program. Programs that had to wait for user input were inherently single process. CPU designers also started to place large amounts of random-access memory (RAM) directly on the CPU chips so that memory fetches occurred faster.

By 2020, Moore's Law had begun to significantly slow as electronic components on integrated circuits were being packed so close that quantum effects limited further progress. Microchips had continued to diversify, with the general-purpose CPU no longer the cutting edge of development. Chips that concentrated on graphics processing and artificial intelligence (AI) processing now led the pack, mainly because those types of programs were often inherently parallel in nature and could take advantage of an ever-increasing number of cores. In 2020, a new massive microchip, almost 72 square inches in size, contained an astounding 2.6 trillion transistors and 850,000 cores. The chip was designed by Cerebras Systems specifically for AI applications.

## MINICOMPUTERS

Between 1965 and 1985, principally, a new class of computers allowed smaller businesses and scientific laboratories to computerize processes at less than a tenth of the cost of a traditional mainframe computer. Computers like the IBM 360 cost, at a minimum, more than $250,000 in 1969. They became known as *mainframes* because of the metal frames that held the processing and memory components. The newer, smaller, and cheaper *minicomputers* were named in the press around 1965 following the naming convention of the miniskirt and Austin Mini.

Ken Olsen (1926–2011) founded Digital Equipment Corporation (DEC) in 1957. Olsen grew up in Connecticut with Scandinavian immigrant parents, served in the navy between 1944 and 1946, and graduated from MIT with a BS and MS in electrical engineering. DEC began manufacturing computers using transistor technology and produced the PDP-1 (Programmed Data Processor-1) computer in 1959. The machine has been celebrated as the platform for the first word processor (Expensive Typewriter) and first video game (*Spacewar!*) in 1961 and 1962, respectively, at MIT. DEC introduced the PDP-8 in 1965. It was the first mass-produced computer based on integrated circuits instead of transistors. The entire PDP-8 fit in a normal packing crate and cost about $18,000.

Minicomputers were initially not as powerful as mainframes and were often bought to be dedicated to a small number of tasks rather than as a general-purpose business data processing computer. The majority sold to original equipment manufacturers (OEMs) that used the machines as embedded processors in industrial control systems. Based on its minicomputers, DEC was the third-largest computer manufacturer in the world in 1970, behind IBM and Sperry Rand. DEC eventually became the world's second-largest computer company, based on the strength of its PDP series and subsequent VAX series of minicomputers.

By the early 1980s, these minicomputers began competing with the lower-end mainframes, thus benefiting smaller companies. Even larger companies realized they could save money by decentralizing computing to separate business units in their organizations. Over 300,000 minicomputers operated in the United States by 1980. DEC owned more than a third, followed by Data General, Honeywell, Hewlett-Packard, Prime Computer, Harris, Apollo, and Wang. Mainframe manufacturers IBM, Honeywell, NCR, and Sperry Rand also entered the space. IBM introduced the System/3 in 1969—a minicomputer based on earlier work by IBM Germany—and the System/38 in 1980, which brought them almost a third of the minicomputer market. During the height of the classic minicomputer market, more than eight-dozen companies, many overseas, created machines. The biggest of the minicomputer companies resided in Massachusetts, frequently on Route 128, a highway that encircled Boston and was often called "America's Technology Superhighway." Politicians called this the "Massachusetts Miracle."

Minicomputers benefited from the introduction of integrated circuits and third-party creation of software and peripherals, which drove down the cost of researching and developing computers, operating systems, and software applications. A number ended up in universities to give computing students access to the inner workings of the machine, an experience they were unlikely to get on a terminal or through batch processing. Minicomputers became more sophisticated. DEC introduced the first supermini in 1978, the 32-bit VAX with a VMS operating system that included virtual memory and multitasking.

When the personal computer emerged in the 1970s, minicomputers occupied the middle ground between microcomputers and mainframes. Minicomputers increasingly ran more sophisticated operating systems, often a variation of UNIX, and were heavily used in the engineering, scientific, academic, and research fields. In the 1980s, minicomputers like Sun Microsystems machines also found their way to the desktop as workstations, powerful single-user machines that were often used for graphics-intensive design and engineering applications.

In the 1990s, minicomputers and workstations disappeared as market segments when personal computers became powerful enough to completely supplant them. Minicomputer companies like DEC attempted to fit into this new paradigm in the 1980s with personal computers such as the DEC Rainbow in 1982. The Massachusetts Miracle was over by 1998 when DEC was sold to Compaq and later to Hewlett-Packard. Most other Route 128 minicomputer companies also folded or were acquired in mergers. The minicomputer had played an important role in the evolution of computing. For example, multitasking and multiuser capabilities became standard features in microcomputer operating systems, such as OS/2 and Windows NT, and different flavors of UNIX became the operating system of choice in microcomputer-based server farms.

## TIME-SHARING

Early computers were all batch systems, meaning a program was loaded into a computer and run to completion before another program was loaded and run. This serial process allowed each program to have exclusive access to the computer but frustrated programmers and users for three reasons. First, the CPU lay idle while programs were loaded, which wasted expensive computer time; second, batch processing made it difficult to do interactive computing; and third, users had to wait for other jobs to finish. Today, you can still find former users of these systems telling stories of how they were moved to the top of the queue by delivering donuts and coffee to the system managers.

In a 1959 memorandum, John McCarthy, already a founding pioneer in artificial intelligence (AI), proposed to MIT that a "time-sharing operator program" be developed for their new IBM 709 computer that IBM planned to donate to the prestigious school. In the United Kingdom, Christopher Strachey (1916–1975) simultaneously and independently came up with the same idea. A prototype of the Compatible Time-Sharing System (CTSS) was running at MIT by late 1961. Further iterations of CTSS were created, and in the mid-1960s, a version of CTSS implemented the first hierarchical file system, which is familiar to users today as the idea of putting files into directories or folders to better organize the file system. J. C. R. Licklider (1915–1990) (someone we shall see later with the invention of the internet) of the Advanced Research Projects Agency (ARPA), a branch of the Pentagon, was a keen advocate of interactive computing and funded continued work on time-sharing. Other American and British projects also researched the goal of getting multiple programs to run in the same computer, a process called *multiprogramming*. Though only one program at a time could actually run on a single-core CPU, other programs could be quickly switched in to run as long as these other programs were also resident in memory. This led to the problem of how to keep multiple programs in memory and not accidentally have one program overwrite or use the memory already occupied by another program. The solution to this was a series of hardware and software innovations to create virtual walls of exclusion between the programs.

Operating system software became much more sophisticated to support multiprogramming and the principle of exclusion. An ARPA-MIT project called Multiplexed Information and Computing Service (or Multics), began in 1965, did not realize its ambitious goals in that it was only a modest commercial success, but it became a proving ground for many important multiprogramming innovations. Two programmers who worked on Multics, Dennis M. Ritchie (1941–2011) and Ken Thompson (1943–) at AT&T's Bell Laboratories, turned their experience into the UNICS operating system. The name stood for Uniplexed Information and Computing Service, a pun on "Multics," but was later shortened to UNIX. Originally written in assembly language on a DEC PDP-7 minicomputer, Ritchie and Thompson wanted to port UNIX to a new minicomputer, the DEC PDP-11, and decided to rewrite the operating system in a higher-level language. Ritchie had created the programming language C (a successor to a language called B), and the rewritten UNIX was the first operating system written in a third-generation language.

As a government-sanctioned monopoly, AT&T was not allowed to sell any of its inventions outside of the telephone business, so AT&T offered UNIX to anyone who wanted to buy it for the cost of the distribution tapes and manuals, though AT&T retained the copyright. Because it was a full-featured operating system with all the source code included, UNIX became popular at universities in the 1970s and 1980s. *The C Programming Language* by Brian Kernighan (1942–) and Ritchie, published in 1978 and updated in 1989, including the American National Standards Institute (ANSI) standardization, became the standard text on the C language. Sometimes affectionately called *K&R*, it is terse but comprehensive, and copies were still being published in 2022.

## IBM SYSTEM/360

In the early 1960s, IBM had seven mutually incompatible computer lines serving different segments of the market. Plans were created for an 8000 series of computers, but a few visionaries within the company argued that creating yet another computer with its own new instruction set would only increase the confusion and escalate manufacturing and support costs. IBM engineers did not even plan to make the different models within the 8000 series computers compatible with each other. Such developments showed that IBM lacked a long-range vision.

Bob O. Evans (1927–2004), an IBM electrical engineer turned manager, led the charge to create a new product line that would supplant the 8000 series and completely replace all the computer systems that IBM manufactured with a uniform architecture of new machines. Frederick P. Brooks Jr. (1931–), who had earned a doctorate in applied mathematics from Harvard University in 1956, was the systems planning manager for the 8000 series and fought against Evans's plans. After the corporation decided to go with the plan proposed by Evans, the canny engineer asked Brooks to become a chief designer of the new system. Gene M. Amdahl (1922–2015), another brilliant young engineer, joined Brooks in designing the System/360.

Honeywell cemented the need for the System/360 computers when its Honeywell 200 computers, introduced in 1963, included a software utility allowing the new Honeywell computer to run programs written for the IBM 1401 computer. The 1401 was a major source of IBM profits, and the cheaper Honeywell 200 threatened to sweep the low-end market for data processing machines.

When Amdahl and Brooks decided that they could no longer work with each other, Evans solved the problem by keeping Amdahl as the main system designer and moving Brooks over to head the difficult task of creating a new operating system for the System/360. The designers of the operating system initially chose to create four different operating systems, for different sizes of machines, to be labeled I, II, III, and IV. This plan, which was based on Roman numerals and did not include compatibility between the different systems, was canceled in early 1964 because it conflicted with the overall design goal of system compatibility. The resulting OS/360 proved to be a difficult challenge, and even after the System/360 was announced in April 1964, the operating system was too full of bugs to be released. Part of the reason that

The IBM System/360 mainframe computer with peripheral devices. IBM Corporate Archives. (Reprint Courtesy of IBM Corporation ©)

the operating system fell behind is that IBM did not charge for software and thus thought of itself primarily as a hardware vendor, not a software vendor. But the OS/360 experience showed that software was becoming more important, and IBM executives paid more attention to software efforts thereafter. The fraction of total research and development efforts at IBM devoted to software rose from one-twentieth in 1960 to one-third in 1969.

Brooks had wanted to retire and go to work at a university, but he remained another year to help work the bugs out of OS/360. His experiences with this project led him to write *The Mythical Man-Month: Essays on Software Engineering* in 1975, which became a classic treatise in the field. A *man-month* is how much work a person can do in a month. If a task is said to take twenty man-months, then one person must work twenty months or ten people must each work two months. As the OS/360 project fell behind, IBM added more programmers to the project, which bloated the size of the teams, making communications between team members more complex and actually increasing the difficulty of completing the project. Brooks compared large programming projects to falling into a tar pit, and he pointed out that programming should be a disciplined activity similar to engineering, with good process control, teamwork, and adherence to good design principles. Brooks also warned against the second system effect, where programmers who disciplined themselves on their first project relaxed and got intellectually lazy on their second project.

In 1965, after spending half a billion dollars on research and another $5 billion on development, IBM shipped the first System/360 machine to a customer. A deluge of orders forced IBM to dramatically expand its manufacturing facilities within a year. By the end of 1966, a thousand System/360 systems were being built and sold every month. The company found that its gamble had paid off, and the company increased its workforce by 50 percent in the next three years to keep up with demand, reaching almost a quarter of a million employees. By the end of the decade, IBM dominated at least 70 percent of the worldwide computer market.

The System/360 achieved its goal of upward and downward compatibility, allowing programs written on one system to run on a larger or a smaller system. Standardized peripheral devices, such as printers, disk drives, and terminals, would work on any of the System/360 machines. By having more uniform equipment, IBM also reined in manufacturing costs. IBM had earlier used the same strategy to dominate the market for punched card machines, making a uniform family of compatible machines that came in different models.

The IBM engineers played it safe with the technology in the System/360, choosing to use solid logic technology (SLT) instead of integrated circuits. SLT used transistors in a ceramic substrate, a new technology that could be mass-produced more quickly. Though IBM advertised the System/360 as a third-generation computer, the technology remained clearly second generation. The System/360 standardized on 8 bits to a word, making the 8-bit byte universal. The System/360 also provided the features necessary to succeed as both a business data processing computer and a number-crunching scientific computer. IBM priced the systems as base systems and then added peripherals and additional features as extra costs. IBM did such a good job of creating standardized computers that some computers were built with additional features already installed, such as a floating-point unit, and shipped to customers with those features turned off by electronic switches. Some customers, especially graduate students at universities, turned on the additional features to take advantage of more capabilities than the university had paid for.

In the interests of getting their project done faster, the OS/360 programmers chose not to include dynamic address translation, which allowed programs to be moved around in memory and formed an important foundation of time-sharing systems. IBM fixed this problem and some of the other technical problems in the System/360 series with its System/370 series, introduced in 1970, by adding dynamic address translation, which became known as *virtual memory*.

The IBM System/360 became so dominant in the industry that other computer manufacturers created their own System/360-compatible machines, such as RCA with its Spectra 70 series, competing with IBM in its own market with better service and cheaper prices. The British ICL 2900 series was System/360 compatible, as was the RYAD computer series built behind the Iron Curtain for Soviet and Eastern European use.

After his instrumental role in designing the IBM 704, IBM 709, and System/360, Gene M. Amdahl grew frustrated that IBM would not build even more powerful machines. IBM based its customer prices proportional to computer processing power, and more powerful computers proved too expensive if IBM retained that pricing model. Amdahl retired from IBM in 1970 and founded his own company, Amdahl Corporation, to successfully build IBM-compatible processors that cost the same but were more powerful and took up less space than comparable IBM machines. Amdahl made clones of IBM mainframes a decade before clones of IBM personal computers completely changed the personal computer market.

## BIRTH OF THE SOFTWARE INDUSTRY

By the mid-1960s, a small but thriving software services industry existed that performed contracts for customers. In 1964, one of these companies, Applied Data Research (ADR), founded in 1959 by seven programmers from Sperry Rand, was approached by the computer manufacturer RCA to write a program to automatically create flowcharts of a program. *Flowcharts* are visual representations of the flow of control logic in a program and are very useful to designing and understanding a program. Many programmers drew flowcharts by hand when they first designed a program, but as the program changed over time, these flowcharts were rarely updated and became less useful, as the changed program no longer resembled the original. After writing a flowcharting program, ADR asked RCA to pay $25,000 for the program. RCA declined the offer, so ADR decided to call the program Autoflow and went to the hundred or so customers of the RCA 501 computer to directly sell the program to them. This was a revolutionary step and resulted in only two customers, who each paid $2,400.

ADR did not give up. Realizing that the RCA market share was too small, the company rewrote Autoflow to run on IBM 1401 computers, the most prevalent computer at the time. The most common programming language on the IBM 1401 was called Autocoder, and Autoflow was designed to analyze Autocoder programming code and create a flowchart. This second version of Autoflow required programmers to insert one-digit markers in their code indicating the type of instruction for each line of code. This limitation was merely an inconvenience if the programmer was writing a new program, but it was a serious impediment if the programmer had to go through old code to add the markers. Customers who sought to buy Autoflow wanted the product to produce flowcharts for old code, not new code, so ADR went back to create yet another version of Autoflow.

This third try found success. Now IBM 1401 customers were interested, though sales were constrained by the culture that IBM had created. Because IBM completely dominated the market and bundled its software and services as part of its hardware prices, independent software providers could not compete with free software from IBM, so they had to find market niches where IBM did not provide software. In the past, if enough customers asked, IBM always wrote a new program to meet that need and gave it away for free. Why should an IBM 1401 customer buy Autoflow when IBM would surely create the same kind of program for free? In fact, IBM already had a flowcharting program called Flowcharter, but it required the programmer to create a separate set of codes to run with Flowcharter and did not examine the actual programming code itself.

Autoflow was clearly a superior product. But executives at ADR recognized the difficulty of competing against free software, so they patented their Autoflow program to prevent IBM from copying it. This led to the first patent issued on software in 1968, a landmark in the history of computer software. A software patent is literally the patenting of an idea that can only be expressed as bits in a computer, not as a physical device, as patents had been in the past.

ADR executives also realized that the company had a second problem. Computer programmers were used to sharing computer code with each other, freely exchanging magnetic tapes and stacks of punched cards. This made sense when software was free and had no legal value, but ADR could not make a profit if its customers turned around and gave Autoflow away to their friends. Because there was no technical way to protect Autoflow from being copied, ADR turned to a legal agreement. Customers signed a three-year lease agreement, acquiring Autoflow like a piece of equipment, which they could use for three years before the lease must be renewed. With the success of the IBM System/360, ADR rewrote Autoflow again to run on the new computer platform. By 1970, several thousand customers used Autoflow, making it the first packaged software product. This success inspired other companies.

The next software product began as a file management system in a small software development company owned by Hughes Dynamics. Three versions, Mark I, Mark II, and Mark III, became increasingly more sophisticated during the early 1960s and ran on IBM 1401 computers. In 1964, Hughes Dynamics decided to get out of the software business, but it had customers who used the Mark series of software and did not want to acquire a bad reputation by just abandoning those customers. John A. Postley (1924–2004), the manager who had championed the Mark software, found another company to take it over. Hughes paid a software services firm called Informatics $38,000 to take its unwanted programmers and software responsibilities.

Postley encouraged Informatics to create a new version, Mark IV, that would run on the new IBM System/360 computers. He estimated that he needed half a million dollars to develop the program. With a scant $2 million in annual revenue, Informatics could not finance such a program, so Postley found five customers willing to put up $100,000 each to pay for developing Mark IV. In 1967, the program was released, selling for $30,000 a copy. More customers were found, and within a year, over $1 million of sales had been recorded, bypassing the success of the Autoflow product.

Informatics chose to lease the software, but for perpetuity rather than a fixed number of years as ADR had chosen to do with Autoflow. This allowed Informatics to collect the entire lease amount up front rather than over the life of the lease, as ADR did. This revision of the leasing model became the standard for the emerging industry of packaged software products. Informatics initially decided to continue to provide upgrades of new features and bug fixes to its customers free of charge, but that changed after four years; the company then began to charge for improvements and fixes to its own program, again setting the standard that the software industry followed after that.

Despite these small stirrings of a software industry, the computer industry was still about selling computer hardware. When the federal government looked at the computer industry, its antitrust lawyers found an industry dominated by one company to the detriment of effective competition. In 1969, under pressure from an impending federal antitrust lawsuit, IBM decided to unbundle its software and services from its hardware and sell them separately, beginning on January 1, 1970. This change created the opportunity for a vigorous community of software and service providers to emerge in the 1970s and directly compete with IBM. Even though IBM planned to unbundle its products, the federal government did file its antitrust lawsuit on the final day of the Johnson presidential administration (in January 1969). The lawsuit lasted for thirteen years and was a continual irritant that distracted IBM management throughout that time. Eventually, the lawsuit disappeared, as it became apparent in the mid-1980s that IBM was losing market share and no longer posed a monopolistic threat.

The effect of the IBM unbundling decision could be seen in the example of software for insurance companies. In 1962, IBM brought out its Consolidated Functions Ordinary (CFO) software suite for the IBM 1401 computer, which handled billing and accounting for the insurance industry. Because large insurance companies created their own software, designing exactly what they needed, the CFO suite was aimed at smaller and medium-sized companies. Since application software was given away for free until 1970, other companies who wished to compete with IBM also had to create application software. Honeywell competed in serving the life insurance industry with its Total Information Processing (TIP) System, which was closely modeled on IBM's CFO software. With the advent of the System/360, IBM brought out its Advanced Life Information System (ALIS) and gave it away to interested customers, though it was never as popular as the CFO product. After unbundling, literally dozens of software companies sprang up to offer insurance software. By 1972, 275 available applications were listed in a software catalog put out by an insurance industry association. Despite the example of insurance software applications, software contractors still dominated the emerging software industry with $650 million in revenue in 1970, as opposed to $70 million in revenue for packaged software products in that same year.

Another type of computer services provider also emerged in the 1960s. In 1962, H. Ross Perot (1930–2019), an IBM salesman, founded Electronic Data Systems (EDS) in Dallas, Texas. The company bought unused time on corporate computers to run the data processing jobs for other companies. EDS did not buy its first computer until 1965, a low-end IBM 1401. EDS grew by developing the concept of what became known as *outsourcing*, which is performing the data processing functions for other companies or organizations. In the late 1960s, the new Great Society programs of Medicare and Medicaid, championed by president and fellow Texan Lyndon B. Johnson, required large amounts of records processing by individual states. EDS grew quickly by contracting with Texas and other states to perform those functions. Further insurance, Social Security, and other government contracts followed, and by the end of the decade, the stock value of EDS had passed $1 billion. Ross Perot ran for president himself as a third-party candidate in 1992 and 1996 but did not win either election.

## BASIC AND STRUCTURED PROGRAMMING

Even with the new third-generation programming languages, such as FORTRAN and COBOL, programming remained the domain of mathematically and technically inclined people. At Dartmouth College, a pair of faculty members and their undergraduate students aimed to change that by developing a system and language for other Dartmouth students to use who were not majors in science or engineering. The Dartmouth team decided on an ambitious project to build both an interactive time-sharing operating system based on using teletype terminals and a new easy-to-use programming language. In 1964, a federal grant allowed Dartmouth to purchase a discounted GE-225 computer. Even before the computer arrived, General Electric (GE) arranged for the Dartmouth team to get time on other GE-225 computers to create their Beginner's All-purpose Symbolic Instruction

Code (BASIC) system. Dartmouth faculty taught BASIC in only two mathematics classes, second-term calculus and finite mathematics, where students were allowed to use an open lab to learn programming.

Clearly based on FORTRAN and ALGOL, BASIC used simple keywords, such as PRINT, NEXT, GOTO, READ, and IF THEN. GE adopted BASIC as its commercial time-sharing system, and within several years, BASIC was ported to computers from other manufacturers. BASIC became the most widely known programming language because of its ease of use and because personal computers in the 1970s and 1980s adopted BASIC as their entry-level and command line (the text prompt with which users communicated with the operating system) language. Early forms of the language were compiled, though the personal computer implementations were usually interpreted. *Compiled programs* are programs that have been run through a compiler to convert the original source code into binary machine code ready to be executed in the CPU. Interpreted code is converted to machine code one line at a time as the program is run, resulting in slower execution. Compiled programs only have to be compiled once, while interpreted programs have to be interpreted every time that they run. Interpreted programs use more computing resources, although they are good for prototyping because they are faster to write and run and obviously better for an operating system interface where you do not want to wait for your commands to compile.

All the early programming languages used some form of "goto" statements to unconditionally transfer control from one section of the program to another section. This method led to what became known as "spaghetti code" because of the programmer's experience while trying to follow the overlapping paths of logic in a program. This problem particularly occurred when programs were modified again and again, with ever more layers of logical paths intertwined with earlier logical paths. Programmers recognized that this was a serious problem but did not know what to do about it.

The Dutch computer scientist Edsger W. Dijkstra (1930–2002) came to the rescue. The son of a chemist and a mathematician, Dijkstra almost starved to death during the famine in the Netherlands at the end of World War II. After obtaining his doctorate with his dissertation exploring "Communication with an Automatic Computer," Dijkstra made a name for himself in the 1950s and 1960s as an innovative creator of algorithms, developing the famous shortest-path algorithm and the shortest spanning tree algorithm. He also contributed work on the development of mutual exclusion to help processes work together in multiprogramming systems.

In 1968, as an eminent programmer, he sent an article to the *Communications of the ACM* journal, "A Case against the GO TO Statement." The editor of the journal, Niklaus Wirth (1934–), chose to publish the article as a letter to the editor to bypass the peer-review process in the journal and speed up its publication. Wirth also picked a more provocative title: "Go To Statement Considered Harmful."

Dijkstra showed that the GOTO statement was actually unnecessary in higher-level languages. Programs could be written without using the GOTO and thus be easier to understand. This insight led to *structured programming*, and newer languages, such as C and Pascal (the latter designed by Wirth), allowed the GOTO to only act within the scope of a function or procedure, thus removing the bad effects of the instruction. Structured programming, the dominant programming paradigm in the 1970s and 1980s, allowed programmers to build larger and more complex systems that exhibited fewer bugs and were easier to maintain. Structured programming was only useful in the higher-level languages, since on the level of machine code, the actual bits that run on a CPU, the GOTO instruction, called a *jump instruction*, is still necessary and pervasive. Structured programming would be followed by a newer approach, *object-oriented programming*, a language approach that made reusability between programs even easier, which began with SIMULA in 1962, created by Kristen Nygaard (1926–) and was made popular in the 1990s with the languages C++, Java, and others.

## SUPERCOMPUTERS

Seymour Cray (1925–1996) showed his passion for electronics as a child, building an automatic telegraph machine at the age of ten in a basement laboratory that his indulgent parents equipped for him. After service in the army during World War II as a radio operator, Cray earned a bachelor's degree in electrical engineering and a master's degree in applied mathematics before entering the new computer industry in 1951. He worked for Engineering Research Associates (ERA), designing electronic devices and computers. When ERA was purchased by Remington Rand (later called Sperry Rand), Cray designed the successful UNIVAC 1103 computer.

A friend left Sperry Rand in 1957 to form Control Data Corporation (CDC). Cray followed him and was allowed to pursue his dream of building a powerful computer for scientific computing. The result was the Control Data 1604 in 1960, which was built for the U.S. Navy. The most powerful computer in the world at that time, it was built entirely of transistors. The new category of *supercomputer* had been born, successors to the IBM Stretch project and the Sperry Rand LARC projects of the late 1950s. Cray continued to design new supercomputers, and the Control Data 6600, released in 1964, included a record 350,000 transistors. Supercomputers were used on the most difficult computing problems, such as modeling weather systems or designing complex electronic systems. Annoyed at the dominance of CDC in the new supercomputer field, IBM engaged in questionable business practices, such as announcing future supercomputer products that they did not ship, that led CDC to file an antitrust suit in 1968. The suit was settled in CDC's favor in 1973.

In 1972, Cray left CDC to found his own company, Cray Research, in his hometown of Chippewa Falls, Wisconsin. CDC generously contributed partial funding to help the new company. Cray was famous for his intense focus and hard work, though he also played hard; besides sports, he allegedly enjoyed digging tunnels by hand on his Wisconsin property to help him think.

In 1976, the Cray-1 was released, costing $8.8 million, with the first model installed at the Los Alamos National Laboratory. Using vector processing, the Cray-1 could perform thirty-two calculations simultaneously. A refrigeration system using Freon dissipated the intense heat generated by the closely packed integrated circuits. Other improved systems followed: the Cray X-MP in 1982, the Cray-2 in 1985, and the Cray Y-MP in 1988. The last machine was the first supercomputer to achieve over a gigaflop in speed (one billion floating-point operations per second); by contrast, the Control Data 6600 in 1964 could only do a single megaflop (one million floating-point operations per second). Every Cray machine pushed the technology envelope, running at ever faster clock speeds and finding new ways of making more than one processor run together in parallel. The name *Cray* was synonymous with supercomputers, though the company's share in the supercomputing market fell in the 1990s as parallel-processing computers from other companies competed to build ever more powerful supercomputers. Cray Research merged with Silicon Graphics Inc. (SGI) in early 1996. Cray died as a result of injuries from an automobile accident later that year.

## MICROPROCESSORS

In 1968, Robert Noyce and Gordon E. Moore decided to leave Fairchild Semiconductor to found Intel Corporation. They raised $500,000 of their own money and obtained another $2,500,000 in commitments from venture capitalists on the basis on their reputations and a single-page proposal letter. Intel had a product available within a year, a 64-bit static random-access memory (RAM) microchip to replace magnetic-core memory. IBM had already created such a technology and used it in its mainframe computers for temporary storage, but did not sell it separately. The Intel microchip crammed about 450 transistors onto the chip.

Intel also introduced dynamic random-access memory (DRAM) technology in 1970, which required a regular electric refreshing on the order of 1,000 times a second to keep the bit values stable. Magnetic-core memories retained their bit values even if the power was turned off, while the new Intel technologies lost everything when power was cut. After only a couple of years, computer system designers adapted to this change because the new memory chips were so much cheaper and faster. Intel also licensed their technology to other microchip manufacturers so that they were not the sole source of the memory chips, knowing that computer manufacturers felt more comfortable having multiple suppliers.

Intel also invented erasable programmable read-only memory (EPROM) microchips in 1970. EPROMs are read-only memory (ROM) chips with a window on top. By shining ultraviolet light into the window, the data on the microchip are erased and new data can be written to the microchip. This technology served the controller industry well, making it easy to embed new programs into their controllers. The EPROM provided a significant portion of Intel's profits until 1984. In that year, the market for microchips crashed; within nine months, the price of an EPROM dropped by 90 percent. Japanese manufacturers had invested heavily in the memory chip market and markets for other kinds of microchips, and manufacturing overcapacity drove prices below any conceivable profit margins. American memory chip manufacturers filed a legal suit alleging illegal dumping by the Japanese. The federal government became involved, and although most American memory chip manufacturers withdrew from the market, the EPROM market was saved. By the mid-1980s, Intel no longer needed the EPROM market because it was then chiefly known as a manufacturer for its fourth major invention: the microprocessor.

In April 1969, Busicom, a Japanese manufacturer of calculators, approached Intel to manufacture a dozen microchips that it had designed for a new electronic calculator. Ted Hoff (1937–), who earned a doctorate in electrical engineering from Stanford University in 1962, was assigned to work with Busicom. Hoff determined that the Japanese design could be consolidated into just five chips. Intel convinced the Japanese engineers to allow them to continue trying to make even more improvements. The Japanese agreed, and Hoff finally got the count down to three: a ROM microchip, a RAM chip, and a microprocessor. The 4-bit microprocessor, called the Intel 4004, contained all the central logic necessary for a computer on a single chip, using about 2,000 transistors on the chip. Stanley Mazor (1941–) helped with programming the microprocessor, and Federico Faggin (1941–) did the actual work in silicon.

By March 1971, the microprocessor had been born. Intel executives recognized the value of the invention, but Busicom had negotiated an agreement giving it the rights to the new microchip. When Busicom experienced financial difficulties, it wanted to negotiate a lower price for the microprocessors. Intel agreed to this lower price as long as Busicom allowed Intel to pay back the $65,000 in research money that Busicom had originally paid to Intel in return for Intel gaining the right to sell the microprocessor to other companies. Busicom agreed, and Intel offered the microprocessor for sale.

While the 4004 was still in development, Hoff designed another microprocessor, the 8-bit Intel 8008. This chip was again developed for an outside company, Computer Terminal Corporation (CTC). When CTC could not buy the microprocessor because of financial difficulties, Intel again turned to selling it to other customers. The Intel 8008 found a role as an embedded data controller and in dedicated word processing computers. The Intel 8008 led to the 8-bit Intel 8080, brought to market in 1974, which became the basis of the first personal computer (PC). The fourth generation of computer hardware was based on microprocessors and ever more densely packed integrated circuits. Intel and other companies sold seventy-five million microprocessors worldwide in just 1979, a strong indication of the outstanding success of Hoff's invention less than a decade later.

By 1960, fewer than 7,000 electronic digital computers had been built worldwide. By 1970, the number of installed electronic digital computer systems stood at about 130,000 machines. Yet, computers remained expensive and were only found in workplace or research settings, not in the home. In the 1970s, the microprocessor became the key technology that enabled the computer to shrink to fit the home.

# FIVE

# Personal Computers: Bringing the Computer into the Home

## THE ALTAIR 8800

When Ted Hoff of Intel created the Intel 4004 microprocessor, a complete central processing unit (CPU), he had created the potential for someone to build a small computer—a microcomputer. However, Intel management wanted to stay out of end user products sold directly to the consumer, so they did not take the next obvious step and create the first microcomputer. The rapid release of Intel's 8008 and 8080 microprocessors soon led a programmer, Gary Kildall (1942–1994), who we explore below, to begin creating a rudimentary operating system for the Intel microprocessors. Kildall and other computer hobbyists had a dream to create a "desktop" computer—a singular computer for their own personal use.

Electronic hobbyists were part of a small community of experimenters who read magazines like *Popular Electronics* and *Radio-Electronics*, attended conventions and club meetings devoted to electronics, and built home electronics systems. They often shared their discoveries with each other. The technical director of *Popular Electronics*, Les Solomon, liked to spur the development of electronic technology by asking for contributions about a particular topic. The submissions that met Solomon's standards would get published in the magazine. In 1973, Solomon put out a call for "the first desktop computer kit." A number of designs were submitted, but they all fell short, until Edward Roberts (1941–2010) contacted Solomon. The cover story of the January 1975 issue introduced the new Altair 8800.

Edward Roberts was born in Miami, Florida. From an early age, he had two primary, seemingly disparate, interests in life: electronics and medicine. Family obligations and financial constraints caused him to pursue electronics. At the time of the Les Solomon challenge, Roberts ran his Micro Instrumentation and Telemetry Systems (MITS) calculator company, one of the first handheld calculator companies in the United States, in Albuquerque, New Mexico. Small companies like his were running into serious competition in the calculator market from big players such as Texas Instruments and Hewlett-Packard. Roberts decided that he would devote his resources to try to meet Solomon's challenge and build a desktop computer with the hope of selling it to hobbyists. He realized that this was a big gamble because no one knew what the market for such a machine might be. He designed and developed the Altair for over a year before he sent a description to Solomon.

The name for the computer has some mythology around it. In a talk, Solomon indicated it came about when Roberts wondered aloud what he should call the machine and his daughter suggested the Altair, as that was the name of the planet that the starship *Enterprise* was visiting that night on *Star Trek*. The episode "Amok Time" does have the *Enterprise* visit Altair 6. Altair is also the planet in *Forbidden Planet*, the planet of the Krell, who had created an almost planet-sized computer and their own doom. Another story indicated that its announcement was stellar enough that the magazine named it after a star. Somewhat ironically, despite the obvious influence of science fiction in the Altair's creation, writers and moviemakers before the 1970s spent little time imagining personal computers in the homes of average people.

The Altair 8800 microcomputer was based on the 8-bit Intel 8080 microprocessor and contained only 256 bytes of memory. The kit cost $397 and came completely unassembled. A person could pay $100 more if they wanted to receive the Altair 8800 already assembled. The microcomputer had no peripherals: no keyboard, computer monitor, disk drive, printer, software, operating system, or any input or output device other than the toggle switches and lights on the front panel of the machine. Programs and data were loaded into memory through the toggle switches, using binary values, and the results of a program run were displayed as binary values on the lights. The Altair was a true general-purpose computer, a von Neumann machine, with the capacity for input and output, even if rudimentary.

Roberts knew that peripheral devices would have to come later. To accommodate integrating them into the machine, the Altair had an open bus architecture. It consisted of a motherboard that held the CPU and expansion slots for cards (circuit boards) that eventually connected to a computer monitor or television, disk drives, or printers. These expansion cards would be built by MITS, hobbyists, and aftermarket companies. Communication between the CPU and the expansion slots occurred through a bus, an electronic roadway by which the CPU checks to see what device on the computer needs attention.

The Altair 8800 microcomputer. (Image Courtesy of the Charles Babbage Institute Archives, University of Minnesota Libraries, Minneapolis)

Four thousand orders for the Altair came to MITS within three months of publication of the *Popular Electronics* article describing the machine, demonstrating a surprisingly large market for home computers. Roberts had trouble obtaining parts, found that parts were not always reliable, and was unprepared to quickly manufacture that many machines, so it took months for MITS to fulfill the orders.

Despite the problems, electronic hobbyists were willing to purchase the microcomputer and put up with long delivery times and other problems. Altair clubs and organizations quickly sprang into existence to address the potential and needs of the machine. Some of these hobbyists became third-party manufacturers and created many of the peripherals needed by the machine, such as memory boards, video cards that could be attached to a television, and tape drives for secondary storage.

Although often given credit for inventing the personal computer (PC), Roberts did not create the first inexpensive desktop computer. In France, André Truong Trong Thi (1936–2005) created and sold a microcomputer called the Micral, based on the Intel 8008, in 1973. Truong sold 500 units of his Micral in France, but the design was never published in the United States. Though the Altair was not first, the size of the electronic hobbyist market in the United States and the open nature of the Altair's design contributed to the speedy development of microcomputers in the United States. All later development of microcomputers sprang from the Altair 8800, not the Micral.

## ORIGINS OF MICROSOFT

Microsoft was started by Paul Allen (1953–2018) and Bill Gates (1955–) and owes its origins to the success of the Altair 8800. Allen was born to librarian parents who inspired his many interests. Gates was born to William Henry Gates Jr., a prominent attorney, and Mary Maxwell, a schoolteacher turned housewife and philanthropist.

Allen and Gates grew up together in Washington State. They were both enthusiastic about computing technology, and Gates learned to program at age thirteen. The enterprising teenagers both worked as programmers for several companies, including automotive parts supplier TRW, without pay, just for the fun of it. While in high school, they created a computer-like device that could measure automotive traffic volume. They called the company Traf-O-Data. The company was short-lived but useful for the two in gaining business experience. Gates may have also created one of the first worms—a program that replicates itself across systems—when he created a program that moved across a network while he was still a junior in high school.

Gates was a student at Harvard University and Allen was working for Honeywell Corporation in Boston when Roberts's *Popular Electronics* article was published. Allen called Roberts in Albuquerque and found MITS had no software for the machine. So Allen called Gates, and they decided to get involved. The two young men were so confident in their technical abilities, and believed that they could draw on the simple BASIC compiler they had already created for the Traf-O-Data machine, that they told Roberts they had a BASIC programming language for the Altair already working. Six weeks later, they demonstrated a limited BASIC interpreter on the Altair 8800 to Roberts. Roberts was sold on the idea and licensed the interpreter from Allen and Gate's newly formed company, Micro-Soft (they later dropped the hyphen). Roberts also hired Allen as his one and only programmer, with the official title of director of software. Gates dropped out of Harvard to help improve the interpreter and build other software for MITS. The BASIC interpreter made operation of the Altair so much easier, opening up the machine to those who did not want to work in esoteric Intel microprocessor machine code.

## MORE MICROCOMPUTERS

MITS was shipping microcomputers out to customers as fast as it could make them, and by the end of 1976, other companies began creating and selling microcomputers as well. A company called IMSAI (Information Management Sciences Associates Incorporated) used the Intel 8080 to create its own microcomputer and soon competed with MITS for market leadership. IMSAI gained some Hollywood fame by appearing in the movie *WarGames* as the microcomputer used by David, the main character played by Matthew Broderick. Companies like Southwest Technical Products and Sphere both used the more powerful Motorola 6800 microprocessor to create their own machines. The company Cromemco developed a computer around the Zilog Z80 chip—a chip designed by former Intel engineer Federico Faggin. MOS Technology, a semiconductor company, created a microcomputer around its 6502 microprocessor and then sold the technology to Commodore and later to Atari. Radio Shack began to look for a machine that it could brand and sell in its stores.

Roberts had not patented the idea of the microcomputer, nor did he patent the idea of the mechanism through which the computer communicated with its components: the bus. Hobbyists and newly formed companies directly copied the Altair bus, standardized it so that hardware peripherals and expansion cards might be compatible between machines, and named it the *S-100 bus*. This meant that engineers could create peripherals and expansion cards for microcomputers that might work in more than just the Altair.

It became obvious to Roberts that the competition was heating up not just for computers but also for the peripherals on his own machine. Most of the profit came from peripherals and expansion cards, so Roberts tried to secure his position by requiring that resellers of the Altair 8800 only sell peripherals and expansion cards from MITS. Most refused to follow his instructions. Manufacturing problems continued as well, and to protect its sales of a problem-plagued 4K memory expansion card, MITS linked the purchase of the card to the popular Micro-Soft BASIC. BASIC normally cost $500, but it only cost $150 if purchased with a MITS memory card. This strategy did not work because a large number of hobbyists simply began making illegal copies of the software and bought memory cards from other manufacturers or made their own.

Seeking a new direction, MITS gambled on the future and released a new Altair based around the Motorola 6800. Unfortunately, hardware and software incompatibility between the new machine and the older 8800 machine, as well as the limited resources MITS had to assign to supporting both machines, did not help MITS in the market. In December 1977, Roberts sold MITS to the Pertec Corporation, and the manufacture of Altairs ended a year later. Roberts left the electronics industry and became a medical doctor; he was able to afford his longtime dream because of the profits from selling MITS. He later went on to combine electronics and medicine, creating a suite of medical laboratory programs in the mid-1990s.

Despite the demise of MITS and the Altair, the microcomputer revolution started by that machine had just begun. Some fifty different companies developed and marketed their own home microcomputers. Many companies would quickly see their own demise. Others were successful for years to come. Commodore introduced its PET in 1977 and followed with even easier-to-use and cheaper models, the VIC-20 and Commodore 64, both based on the MOS 6502 microprocessor. Atari introduced its 400 and 800 machines, also based on the 6502 microprocessor, in 1979. Radio Shack began to sell its TRS-80 (referred to in slang as the "TRASH-80"), based on the Zilog Z80 processor, in its stores nationally in 1977, helping to introduce computing to non-hobbyists.

## THE APPLE II

The genesis of the Apple Computer is found in Homestead High School in Sunnyvale, California, where students were often children of the numerous computer engineers and programmers who lived and worked in the area. Many of these children showed interest in electronic technology, including Steve Wozniak (1950–), often known just by his nickname: "Woz." One of Woz's first electronic devices simulated the ticking of a bomb. He placed it in a friend's school locker as a practical joke. The principal of the school found the device before the friend did and suspended Woz—although just for two days in those more lenient times.

By 1971, Woz had graduated and was working a summer job between his first and second years of college when he began to build a computer with an old school friend, Bill Fernandez (1954–). They called it the Cream Soda Computer because of the late nights they had spent building it while drinking the beverage. By Woz's account, the machine worked, but when they tried to show it to a local newspaper reporter, a faulty power supply caused it to burn up. This story shows how hobbyists were working to create the microcomputer, as the Cream Soda Computer's inauspicious debut came two years before the debut of the Altair 8800. Fernandez also introduced Woz to Steve Jobs (1955–2011).

Jobs was another Silicon Valley student and, by most accounts, a bright, enterprising, and persuasive young man. He once called William Hewlett (1913–2001), one of the founders of Hewlett-Packard, and convinced Hewlett to lend him spare electronics parts. Jobs was twelve years old at the time. Though Jobs was five years younger than Woz when they met, they shared a common affection for practical jokes, and the two got on well. One of their first enterprises together proved rather dubious. They constructed "blue boxes," an illegal device that allowed an individual to make free phone calls, and sold them to their friends. Jobs also obtained a summer job at Atari, a video game company newly founded in 1971 by Nolan Bushnell (1943–). Jobs enlisted Woz to help him program a game that Bushnell had proposed, even though Woz was already working full time at Hewlett-Packard. The game, *Breakout*, became an arcade hit.

Woz began working on another microcomputer in 1975. It was not a commercial product and was never intended to be, just a single circuit board in an open wooden box. Jobs, however, saw commercial potential and convinced Woz that it had a future. They called it the Apple I. As pranksters fond of practical jokes, they decided to found the company on April Fool's Day in 1976. The price of the machine was $666.66. Woz's design, with its organized visible electronic arrangement, was considered "beautiful" by hobbyists.

Jobs's ambition went beyond the handmade Apple I. After consulting with Bushnell, he decided to seek venture capital. He was introduced to Mike Markkula (1942–), a former marketing manager at Fairchild and Intel who had turned venture capitalist. Markkula became convinced that the company could succeed. He secured $300,000 in funding from his own sources and a line of credit.

The Apple II, designed by Woz and based on the MOS 6502 microprocessor, was introduced in 1977. The Apple II cost $790 with 4 kilobytes of RAM or $1,795 with 48 kilobytes of RAM. The company made a profit by the end of the year as production doubled every three months. Though Apple was hiring and bringing in money, Woz continued to work full time at Hewlett-Packard, requiring Jobs to turn his arts of persuasion on Woz to convince him to come work at Apple full time.

The Apple II came in a plastic case that contained the power supply and keyboard. It had color graphics, and its operating system included a BASIC interpreter that Woz had written. With a simple adapter, the Apple II hooked up to a television set as its monitor. The Apple II was an attractive and relatively reliable machine. Many elementary and secondary schools purchased the Apple II across the United States, making it the first computer that many students came in contact with. The microcomputer's open design allowed third-party hardware manufacturers to build peripherals and expansion cards. For example, one expansion card allowed the Apple II to display eighty columns of both upper- and lowercase characters, instead of the original forty columns of only uppercase characters. Programming the Apple II was fairly simple, and many third-party software products were created for it as well. This ease of programming allowed programmers to translate the most popular programs on other hobbyist microcomputers—games like *MicroChess*, *Breakout*, *Space Quarks*, and *Adventure*—which helped the Apple II reach broad acceptance. Many millennials remember this machine as their first computer. However, customers did not think of microcomputers like the Apple as business machines. A program named VisiCalc changed that.

VisiCalc (short for visible calculator) was the first spreadsheet program on a microcomputer. The first electronic spreadsheet on a mainframe computer was LANPAR (LANguage for Programming Arrays at Random) in 1969 at the Massachusetts Institute of Technology (MIT). A spreadsheet is a simple table of cells in columns and rows. The columns and rows go beyond the boundaries of the screen and can be scrolled to either up and down or left and right. Cells may contain text, numbers, or equations that can summarize and calculate values based on the contents of other cells. The spreadsheet emulates a paper accounting sheet but is far more powerful because it can change the value of cells dynamically as other cells are modified. The idea had been around on paper since the 1930s as a financial analysis tool (the double-entry accounting sheet itself was invented in the fifteenth century by Luca Pacioli), but the computer made it a truly powerful idea.

Dan Bricklin (1951–) and Bob Frankston (1949–), two Harvard MBA students, wrote VisiCalc in Frankston's attic on an Apple II connected to a Multics time-sharing system in the 6502 assembly language. They released their program in October 1979 and were selling 500 copies a month by the end of the year. A little more than a year later, VisiCalc was selling 12,000 copies a month at $150 per copy. Users could add functionality to the program using BASIC.

Other powerful business programs were introduced as well. For example, John Draper (1943–), a former hacker known as "Captain Crunch," wrote EasyWriter, the first word processing application for the Apple II. Compared to all other programs, however, VisiCalc was so successful because it drove people to purchase the Apple II just to run it. A new term described this kind of marketing wonder software: the *killer app*. A killer app (or killer application) is a program that substantially increased the popularity of the hardware it ran on. Apple continued to prosper, and in 1981, the company had sales of $300 million a year and employed 1,500 people.

## THE IBM PC

On August 12, 1981, a new player joined the ranks of microcomputer manufacturers: Big Blue. IBM saw the possibility of using microcomputers on business desks and decided it needed to get on the ground with its own microcomputer—and quickly. IBM's intention was to dominate the microcomputer market the same way it dominated the mainframe marketplace, though it anticipated that the microcomputer market would remain much smaller than the mainframe market.

IBM approached the problem of going to market with a microcomputer differently than it had for any other hardware it had produced. The company chose not to build its own chipset for the machine, like it had for its mainframes and minicomputers, and so the new microcomputer used the 16-bit Intel 8088, a chip used in many other microcomputers. IBM also learned from the successes of the Altair 8800 and other microcomputer pioneers by recognizing that IBM needed the many talents of the microcomputer world to build the peripherals and software for its PC. It also decided to go outside IBM for the software for the machine, including the operating system. To facilitate third-party programming and hardware construction, IBM did a few other things that never would have occurred in the mature market of mainframes. IBM created robust and approachable documentation and an open bus-type hardware architecture similar to that of the Altair's S-100 bus. Recognizing the change in the market landscape, IBM also sold the machine through retail outlets instead of only through its established commercial sales force.

Searching for applications for its microcomputer, IBM contacted Microsoft in 1980 and arranged a meeting with Gates and his new business manager, Steve Ballmer (1956–), in Microsoft's Seattle-area offices. Gates's mother may have played a role in Microsoft's eventual overwhelming success. She sat on the board of the United Way with a major executive at IBM, and he recognized Microsoft as her son's business. Gates and Ballmer put off a meeting with Atari to meet with IBM. Atari was in the process of introducing computers for the home market based around the MOS 6502 microprocessor. For the meeting with IBM, Gates and Ballmer decided to look as serious as possible and put on suits and ties—a first for them in the microcomputer business. In another first, they signed a confidentiality agreement so that both Microsoft and IBM would be protected in future development. Microsoft expressed interest in providing software applications for the new machine.

The IBM PC 5150. IBM Corporate Archives. (Reprint Courtesy of IBM Corporation ©)

IBM also needed an operating system and went to meet with Gary Kildall at Digital Research Incorporated (DRI). Kildall had written an operating system called Control Program for Microcomputers (CP/M) that worked on most 8-bit microprocessors, as long as they had 16 kilobytes of memory and a floppy disk drive. This popular operating system ran on the IMSAI and other Altair-like computers, and by 1981 it sat on over 200,000 machines with possibly thousands of different hardware configurations. Before CP/M, the closest thing to an operating system on the microcomputer had been various versions of BASIC. CP/M was much more powerful and could work with any application designed for the machines. However, IBM hesitated at paying $10 for each copy of CP/M. IBM wanted to buy the operating system outright at $250,000. Talking again with Gates, they became convinced that they might be better off with a whole new operating system because CP/M was an 8-bit operating system and the 8088 was 16-bit CPU. So, despite Microsoft not actually owning an operating system at the time, IBM chose Microsoft to provide its microcomputer operating system. Microsoft was a small company among small software companies, bringing in only $8 million in revenue in 1980, when VisiCalc brought in $40 million in revenue in the same year.

Microsoft purchased a reverse-engineered version of CP/M from Seattle Computer Products called SCP-DOS, which it reworked into the Microsoft Disk Operating System (MS-DOS), which IBM called the Personal Computer Disk Operating System (PC-DOS), to run on the Intel 8088 microprocessor. CP/M and MS-DOS not only shared the same commands for the user but also the same internal system calls for the programmers. Kildall considered a lawsuit at this brazen example of intellectual property theft, but he instead reached an agreement with IBM for the large company to offer his operating system as well as the Microsoft version. Unfortunately, when the product came out, IBM offered PC-DOS at $40 and CP/M-86 at $240. Not many buyers went for the more expensive operating system.

A mantra had existed in the computer world for many years: "No one ever got fired for buying IBM." With IBM now in the microcomputer market, businesses that never considered buying a microcomputer prior to the IBM PC were in the market for them. With the introduction of the IBM PC, microcomputers were now referred to generically as personal computers, or PCs, and were suddenly a lot more respectable than they had been. Another killer application appeared that also drove this perception. A program called Lotus 1-2-3, based on the same spreadsheet principle as

VisiCalc, pushed the PC. The introduction of the program included a huge marketing blitz with full-page ads in the *Wall Street Journal.*

For a couple of years in the early 1980s, it was not clear where the microcomputer market would go. Osborne Computer, founded in 1981, created the first really portable PC. The Osborne 1 used a scrollable five-inch screen; contained a Zilog Z80 microprocessor, 64 kilobytes of RAM, and two floppy disk drives; and was designed to fit under an airplane seat. The portable ran CP/M, BASIC, the WordStar word processing software, and the SuperCalc spreadsheet. The Osborne 1 sold for only $1,795, and soon customers were buying 10,000 units a month. Other portable personal computers, almost identical to the Osborne, quickly followed from Kaypro and other manufacturers. In 1980 and 1981, other large computer manufacturers began to bring out PCs, such as Hewlett-Packard, with its HP-85; Xerox, with its Star; and Digital Equipment Corporation (DEC), with its Rainbow (a dual-processor machine that could run both 8-bit and 16-bit software).

Despite the other efforts, the successful combination of IBM and Microsoft eliminated most of the rest of the PC market. By the end of 1983, IMSAI was gone, Osborne had declared bankruptcy, and most of the 300 computer companies that had sprung up to create microcomputers that were not compatible with IBM PCs had disappeared. Kildall's DRI also began its downward spiral as CP/M became less important. By 1983, it looked like there would soon be only two companies left selling microcomputers on a large scale to battle for supremacy: IBM and Apple. *Time* magazine also noticed the importance of the microcomputer when it chose the PC as its Man of the Year for 1982, the only time that it chose a machine for an honor that usually went to an important international leader.

## XEROX PARC, THE GUI, AND THE MACINTOSH

With the introduction of the IBM PC and Microsoft DOS, Apple faced serious competition for the first time, and Jobs turned to formulating a response. As an operating system, DOS adequately controlled the machine's facilities, but few would call the user interface intuitive. Users typed in cryptic commands at the command line to get the machine to do anything. Jobs visited the Xerox Palo Alto Research Center (PARC) in 1979 and came away with a whole different idea for a user interface.

Established in 1970, Xerox PARC was initially headed by Robert "Bob" Taylor (1932–2017), previously the director of the Information Processing

Techniques Office at the Advanced Research Projects Agency (ARPA) in the Pentagon. Taylor helped lay the groundwork for the national network that became the internet and brought his skill at putting together talented people and resources to PARC. Scientists and engineers at PARC quickly established themselves as being on the cutting edge of computing science. PARC created a computer in 1973 called the Alto that used a bitmapped graphical display, a graphical user interface (GUI), a mouse, and programs based on the "what you see is what you get" (WYSIWYG) principle. The GUI used two- or three-dimensional graphics and pointing mechanisms to the graphics, with menus and icons, as opposed to the old method of using text commands at an operating system command prompt. Jobs also saw an Ethernet network linking the computers on different engineers' desks to each other and to laser printers for printing sharp graphics and text.

Though structured programming had only truly started to be established in the industry in the 1970s, programmers at PARC needed more. So Alan Kay (1940–) and the team created Smalltalk, an object-oriented programming (OOP) language better suited for writing a GUI and other graphical programs. Variants of the SIMULA languages designed at the Norwegian Computing Center in Oslo, Norway, in the 1960s were the first examples of OOP, but Smalltalk came to be considered the purest expression of the idea. Structured programs usually separated the data to be processed from the programming code that did the actual processing. Object-oriented programs combined data and programming code into objects, making it easier to create objects that could be reused in other programs. Object-oriented programming utilized structured programming techniques but also required thinking in a different paradigm, and OOP did not gain widespread acceptance until the late 1980s.

With this plethora of riches, practically every major innovation that would drive the computer industry for the next decade, generating hundreds of billions of dollars in revenue, Xerox remained a copier company in its heart. Xerox introduced the 8010 "Star" Information System in 1981, a commercial version of the Alto, but priced it so high at $40,000 that a system and peripherals cost about as much as a minicomputer. Though about 2,000 Star systems were built and sold, this was a failure compared to what Jobs eventually did with the concepts. Because of the failure of Xerox to exploit its innovations, scientists and engineers began to leave PARC to found their own successful companies or to find other opportunities. Jobs eventually convinced a number of the engineers at Xerox PARC to come over to Apple Computer.

Many of Xerox's innovations implemented the prior ideas of Douglas C. Engelbart (1925–2013), a visionary inventor. Raised on a farm, Engelbart entered Oregon State College in 1942, majoring in electrical engineering, under a military deferment program during World War II. After two years, the military ended the deferment program because of a more immediate need for combat personnel versus a longer term need for engineers. Engelbart elected to join the navy and became a technician, learning about radios, radar, sonar, teletypes, and other electronic equipment. He missed the fighting and returned to college in 1946. Two years later, he graduated and went to work for the National Advisory Committee for Aeronautics (NACA), a precursor to the National Aeronautics and Space Administration (NASA). He married in 1951, and feeling dissatisfied with his work at NACA, he sought a new direction in his life.

After considerable study, Engelbart realized that the amount of information was growing so fast that people needed a way to organize and cope with the flood, and computers were the answer. Engelbart was also inspired by the seminal 1945 article "As We May Think," by the electrical engineer Vannevar E. Bush, who directed the American Office of Scientific Research and Development during World War II. Bush had organized the creation of a mechanical differential analyzer before the war, and after the war, he envisioned the use of computers to organize information in a linked manner that we now recognize as an early vision of hypertext. In 1951, electronic computers were in their infancy, with only a few dozen in existence. Engelbart entered the University of California, Berkeley, and earned a master's degree in 1952 and a PhD in electrical engineering in 1955, with a specialty in computers.

Engelbart became an employee at the Stanford Research Institute (SRI) in 1957, and his paper "Augmenting Human Intellect: A Conceptual Framework" laid out many of the concepts in human-computer interaction he had been working on. He formed his own laboratory at SRI in 1963, the Augmentation Research Center (ARC). Engelbart's team of engineers and psychologists worked through the 1960s on realizing his dream, the NLS (oNLine System). Engelbart wanted to do more than automate previous tasks like typing or clerical work; he wanted to use the computer to fundamentally alter the way that people think—to essentially extend the capabilities of human beings. In a demonstration of the NLS at the Fall Joint Computer Conference in December 1968, Engelbart showed the audience onscreen video conferencing with another person back at SRI, thirty miles away; an early form of hypertext; the use of windows on the screen; mixed graphics-text files; structured document files; and the first mouse. This influential technology demonstration became known as "the mother of all demos."

While often just credited with inventing the mouse, Engelbart had also developed the basic concepts of groupware and networked computing. His innovations were ahead of their time and required expensive equipment that retarded his ability to innovate. One of Engelbart's computers at SRI became the second computer to join the ARPANET in 1969, an obvious expansion of his emphasis on networking. ARPANET later evolved into the internet. (ARPA was renamed DARPA in 1972, with "Defense" added in front of the name, then renamed back to ARPA from 1993 to 1996 before reverting back to DARPA. We use the name ARPANET, even though DARPANET was sometimes used during the name changes.) In the early 1970s, several members of Engelbart's team left to join the newly created PARC, where ample funding led to rapid further development of Engelbart's ideas. It only remained for Steve Jobs and Apple to bring the work of Engelbart and PARC to commercial fruition.

Steve Jobs said of his Apple I, "We didn't do three years of research and come up with this concept. What we did was follow our own instincts and construct a computer that was what we wanted." Jobs's next foray into computer development used the same approach. The first attempt by Apple to create a microcomputer that used the GUI interface was the Lisa (named after Jobs's daughter), which was based on a 16-bit Motorola 68000 microprocessor and released in 1983. The Lisa was expensive and noncompatible with both the Apple II computer line and the rest of the DOS-oriented microcomputer market and did not sell well.

After Jobs became disenchanted with the Lisa team during production, he decided to create a small "skunk works" team to produce a similar but less expensive machine. Pushed by Jobs, the team built a computer and small screen combination in a tan box together with a keyboard and mouse: the Apple Macintosh. Also based on the Motorola 68000 microprocessor, the Macintosh was the first successful mass-produced GUI computer.

The Macintosh's public unveiling was dramatic. During the 1984 Super Bowl television broadcast, a commercial flickered on that showed people clothed in gray trudging like zombies into a large, bleak auditorium. A huge television in the front of the auditorium displayed a talking head, similar to the Big Brother character from George Orwell's novel *1984*, droning on. An athletic and colorfully clothed woman who is being chased by characters that look like security forces runs into the room. She swings a sledgehammer into the television. The television explodes, blowing a strong dusty wind at the seated people. A message comes on the screen:

On January 24th,
Apple Computer will introduce
Macintosh.
And you'll see why 1984
won't be like "1984."

The reference to Orwell's novel, where Big Brother is an almost omnipotent authoritarian power, was intriguing. Although never stated, it was not hard to guess that Apple was likening Apple's nemesis, IBM, to Big Brother.

The Macintosh (or "Mac," as it was affectionately called) quickly garnered a lot of attention. Sales were initially stymied by hardware limitations, since the Mac had no hard drive and limited memory and lacked extensive software. Eventually, Apple overcame these initial limitations, allowing the machine to fulfill its promise. Even with the initial problems, the Macintosh suddenly changed the competitive landscape. The development of Aldus PageMaker by Paul Brainerd (1947–) in 1985, the first desktop publishing program, became the killer app for the Macintosh, making it a successful commercial product, just as VisiCalc had made microcomputers into useful business tools.

Engelbart's contributions were lost in popular memory for a time, even at Apple, which at one point claimed in a famous 1980s lawsuit against Microsoft to have effectively invented the GUI. Yet, by the mid-1980s, people in the computer industry began to take notice of Engelbart's contributions, and the awards began to flow. Among his numerous awards were a lifetime achievement award in 1986 from *PC Magazine*; a 1990 ACM Software System Award; the 1993 IEEE Pioneer Award; the 1997 Lemelson-MIT Prize, with its $500,000 stipend; and the National Medal of Technology in 2000.

Jobs was forced out of Apple in 1985 by John Scully (1939–), the man he had handpicked to be the new head of Apple. Jobs reacted by founding NeXT Inc., where he intended to build the next generation of personal computers. The NeXT computer, introduced in 1988, was designed around a 32-bit Motorola 68030 microprocessor, contained 8 megabytes of memory, and included a 256-megabyte magneto-optical drive for secondary storage instead of a floppy disk drive. The NeXT ran a sophisticated variant of the UNIX operating system and included many tools for object-oriented programming, which impressed technical people, but the software ran slowly. The window for introducing a completely new microcomputer architecture had apparently passed for a time. Sales were poor, improved models did not flourish, and the company lost money. The NeXT did gain fame later as the machine on which Tim Berners-Lee (1955–) invented the World Wide Web.

Jobs also cofounded Pixar Animation Studios after purchasing the computer graphics division of Lucasfilm, made famous by the *Star Wars* movies. Pixar created computer-animated movies using proprietary software technology that it developed, and by concentrating on storylines, not deadlines, it began a string of successes with the full-length feature film *Toy Story* in 1995. In 1996, the board of directors of Jobs's old company, Apple, after suffering business losses, asked Jobs to return to head the company. He did so on the condition that Apple would buy NeXT Inc., which it did. The NeXT operating system and programming tools were integrated into the Apple Macintosh line, becoming Mac OS X in 2001. Jobs successfully turned Apple around and relished a sense of vindication.

## IBM PC CLONES

When IBM first approached Microsoft, Bill Gates successfully convinced IBM that its PC should follow the direction of open architecture that it had begun in its hardware by having its PC be able to support any operating system. He pointed to the success of VisiCalc, where software drove hardware sales. Gates figured that he could successfully compete with any other operating system. In many ways, this was not a large gamble. Gates understood that a paradigm-shifting operating system might come along to supplant DOS, but he also knew from his experiences in the microcomputer world that users tended to stick to a system once they had acquired experience with it. This was known as *technological lock-in.*

Gates also argued to IBM that because Microsoft was at risk for potentially having its operating system on the PC replaced by a competitor's, Microsoft should be free to sell its operating system to other hardware manufacturers. IBM bought the argument and opened the door for clones. Gates was acutely aware of the experience of the Altair with its open architecture, which quickly led to clones. The open architecture of the PC meant that third parties could also clone the IBM PC's hardware.

While Apple kept its eye on the feared giant of IBM, other companies grabbed market share from both Apple and IBM by creating IBM PC clones. IBM PCs could be cloned for three reasons: (1) Intel could sell its microprocessors to other companies, not just IBM; (2) Microsoft could also sell its operating system to the clone makers; and (3) the Read-Only Memory Basic Input/Output System (ROM BIOS) chips that IBM developed could be reverse engineered. ROM BIOS chips were memory microchips containing the basic programming code to communicate with peripheral devices, like the keyboard, display screen, and disk drives. A clone market for Apple Computer could not emerge because Apple kept a tight legal hold on its Macintosh ROM BIOS chips, which could not easily be reverse engineered.

One of the first clone makers was also one of the most successful. Compaq Computer was founded in 1982 and quickly produced a portable computer that was also an IBM PC clone. When the company began to sell its portable computers, its first year set a business record, selling 53,000 computers for $111 million in revenue. Compaq moved on to building desktop IBM PC clones and continued to set business records. By 1988, Compaq was selling more than $1 billion of computers a year. The efforts of Compaq, Dell, Gateway, Toshiba, and other clone makers continually drove down IBM's market share during the 1980s and 1990s. The clone makers produced cheaper microcomputers with more power and features than the less nimble IBM.

The majority of PCs sold by 1987 were based on Intel or Intel-compatible microchips. Apple Macintoshes retreated into a fractional market share, firmly entrenched in the graphics and publishing industries, while personal computers lines from Atari and Commodore faded away in 1992 and 1994, respectively. The success of the clone makers meant that the terms *personal computer* and *PC* eventually came to mean a microcomputer using an Intel microprocessor and a Microsoft operating system, not just an IBM personal computer.

Intel saw the advantages of the PC market and continued to push its microprocessors along the path of Moore's Law. The 8088 was a hybrid 8/16-bit microprocessor with about 29,000 transistors. The 16-bit Intel 80286 microprocessor, introduced in 1982, had 130,000 transistors. The 32-bit Intel 80386 microprocessor, introduced in 1985, had 275,000 transistors. The 32-bit Intel Pentium microprocessor, introduced in 1993, contained 3.1 million transistors. The Pentium Pro, introduced in 1995, contained 5.5 million transistors; the Pentium II, 7.5 million transistors; and the Pentium III, released in 1999, had 9.5 million transistors. The Pentium IV, introduced in 2000, used a different technological approach and reached 42 million transistors on a single microchip. The Advanced Micro Devices (AMD) EPYC Rome microprocessors had more than 39 billion transistors in 2019.

One of the major reasons for the success of the Intel-based PC was that other companies also made Intel-compatible chips, forcing Intel to continually strive to improve its products and keep prices competitive. Without this price pressure, PCs would have certainly remained more expensive. In the early 1980s, at the urging of IBM, Intel had licensed its microprocessor designs to other computer chip manufacturers so that IBM might have a second source to buy microprocessors from if Intel factories could not keep up with demand. In the 1990s, after Intel moved away from licensing its products, only one competitor, AMD, continued to keep the marketplace competitive. AMD did this by moving from just licensing Intel technology to reverse engineering Intel microprocessors and creating its own versions. By the early 2000s, AMD was designing different features into the microprocessors than were found in comparable Intel microprocessors and still remaining mostly compatible.

## SOFTWARE INDUSTRY

After IBM's unbundling decision in 1969, which led to the sale of software and hardware separately, the software industry grew rapidly. In 1970, total sales of software by U.S. software firms was less than half a billion dollars. By 1980, U.S. software sales had reached $2 billion. Most of the sales in the 1970s were in the minicomputer and mainframe computer markets. Sales of software for personal computers completely revolutionized the software industry, dramatically driving up sales during the 1980s. In 1982, total sales of software in the United States reached $10 billion; in 1985, it was $25 billion. The United States dominated the new software industry, which thrived in a rough-and-tumble entrepreneurial atmosphere.

Creators of PC software often did not come from the older software industry. They sprang out of the hobbyist and computer games communities and sold their software like consumer electronics products, in retail stores and through hobbyist magazines, not as a capital product with salespeople in suits visiting companies. Hobbyists and gamers also demanded software that was easier to learn and easier to use for their personal computers than the business software that was found on mainframes. This emphasis on human factors design became an important part of the software industry and eventually even affected how mainframe business software was designed.

Around 1982, as the IBM PC and its clones became dominant in the marketplace, the software market became more difficult for the young hobbyist to enter. VisiCalc contained about 10,000 lines of programming code, something that a pair of programmers could easily manage, whereas Lotus 1-2-3, the product that pushed VisiCalc out of the market, contained about 400,000 lines of code, which required a team effort. VisiCalc had sold about 700,000 copies since its launch, whereas Lotus 1-2-3, propelled by $2.5 million in advertising, sold 850,000 copies in its first eighteen months.

As the cost of entering the software marketplace went up, an interesting alternative marketing model emerged. Beginning in about 1983, programmers who created a useful program often offered it to other people as shareware. This usually meant that anyone who wanted to could use the program, and a donation was requested if the program proved useful. Among the more useful programs distributed under this scheme were a word processor, PC-Write; a database, PC-File; and a modem control program, PC-Talk. Many minor games were also distributed as shareware.

## GAMES

By 1982, annual U.S. sales of computer games stood at $1.2 billion. Computer games had their origin in mechanical pinball machines. The first electric pinball machine was built in 1933, and electronics were later included to make the machines more sophisticated and flashier. The first true computer game was invented by MIT graduate student Steve Russell (1937–) in 1962 on a Digital Equipment Corporation (DEC) PDP-1. *Spacewar!* graphically simulated two spaceships maneuvering and firing rocket-propelled torpedoes at each other. Using toggle switches, the users could change both the speed and direction of their ships and fire their torpedoes. Other students added accurate stars for the background and a sun with a gravity field that correctly influenced the motion of the spaceships. The students also constructed their own remote controllers so that their elbows did not grow tired from using the toggle switches on the PDP-1. MIT, Stanford University, and the University of Utah were all pioneers in computer graphics and some of the few places in the early 1960s where a programmer could actually use a video terminal to interact with a computer.

Nolan Bushnell, educated at the University of Utah, played *Spacewar!* incessantly at the university, inspiring him to write his own computer games while in school. After graduating, Bushnell designed an arcade version of *Spacewar!*, called *Computer Space*, and found a partner willing to manufacture 1,500 copies of the game for the same customers who purchased pinball machines, jukeboxes, and other coin-operated machines. Far too complex for amateurs to play, the game failed to sell.

Bushnell did not give up. He partnered with a fellow engineer to found a company called Atari in 1972. While Bushnell worked on creating a multiplayer version of *Computer Space*, he hired an engineer and assigned him to create a simple version of ping-pong that could be played on a television set. *Pong* became a successful arcade video game. Then, in 1975, Bushnell partnered with Sears to sell a version called *Home Pong* in its stores. *Home Pong* attached to the television set at home. The game sold wildly. In 1977, Atari released its Atari 2600, a home unit that could play many games that each came on a separate cartridge. Though Bushnell had been forced out of the company in 1978, his dream of a commercially successful version of *Spacewar!* was realized in 1979 when Atari released *Asteroids*, which became its all-time best-selling game.

Other companies also competed in the home video market, but Atari defined the home video game market in the eyes of many, until the company took awful losses in 1983 as the market for home game consoles crashed. Part of the reason for the crash was that PC games were becoming more popular. Nintendo revived the game console market in 1986, and both Sega and Sony joined the competition—all of them Japanese companies. Nintendo had learned from the mistakes of Atari and kept tight legal and technical control over the prices of game cartridges so that excessive competition would not drive the prices of games down so far that profit disappeared. In the 1990s, game console systems and games for PCs became so popular that the revenue in the game market surpassed the revenue generated by Hollywood movies.

Games for PCs had existed from the start of the PC revolution, but they did not become a powerful market force until about the time that game consoles stumbled in 1983. The PC, with a keyboard, provided a better interface for more sophisticated games rather than just straight arcade-style games. Games such as *Adventure*, from Adventure International (founded 1978); *Zork*, from Infocom (founded 1979); *Lode Runner*, from Broderbund (founded 1980); and *Frogger*, from Sierra Online (founded 1980) define the memories of many PC users of that period.

In the late 1970s, games called multiuser dungeons (MUDs) appeared in Britain and the United States. The games were not created for commercial sale, but for fun, and ran on early networks and Bulletin Board Systems (BBS). Players used a text interface to make their way through dungeons, fight monsters, and interact with other players.

In 1997, *Ultima Online*, a massively multiplayer online role-playing game (MMORPG), showed a new direction for gaming by combining the graphical power and sophistication of single-user PC games with the versatility and multiplayer challenge of MUDs. Later online games, such as *EverQuest* and *World of Warcraft*, successfully followed *Ultima Online*. South Korea, because of its heavily urban population, had over 70 percent of all households connected to the internet via high-speed broadband connections in the early 2000s. The online game *Lineage*, released in 1998, became so popular in South Korea that by 2003 nearly two million people played it every month, out of a total population of less than forty-nine million. *Lineage* was a medieval fantasy epic, which seemed to be the preferred format for successful online games, though that particular game later declined in popularity.

## MICROSOFT ASCENDENT

From its infant beginnings of offering BASIC on the Altair and other early microcomputers, Microsoft grew quickly as its executives effectively took advantage of the opportunities that the IBM PC offered. Microsoft actively aided the growth of the PC clone market; every IBM PC and PC clone required an operating system, which enabled Microsoft to earn revenue on every PC sold. Microsoft also created a single game, *Flight Simulator*, first released in 1983, that was so demanding of the PC's hardware and software that running the game became a litmus test as to whether a new clone model was truly compatible enough with the PCs from IBM.

Over 100 million copies of DOS were eventually sold. Using the revenue from its dominant operating system, Microsoft developed further versions of DOS and funded the development of other software packages. The original DOS 1.0 contained only 4,000 lines of programming code. DOS 2.0, released in 1983, contained five times that much code, and DOS 3.0, released in 1984, doubled the amount of code again, reaching 40,000 lines. From early on, Microsoft developed well-regarded compilers and other programming tools. It also developed other types of application software, such as word processors and spreadsheets, but it was not as successful in those product categories until the 1990s.

Microsoft saw the advantage of the GUI that Engelbart and PARC had invented and hired several top programmers from PARC. Microsoft developed early applications for the Apple Macintosh, even though Apple was a competitor of the IBM PC, and Microsoft also created its own GUI for DOS, called Windows. The first version shipped in 1985, and Microsoft soon followed that with a second version. Both versions were truly awful products: slow, aesthetically ugly, and mostly useless, except for a few programs written to use them.

DOS was a primitive operating system at best, unable to effectively multitask or even effectively manage memory above a 640-kilobyte limit. IBM and Microsoft decided to jointly create OS/2, a next-generation operating system for the PC that would include multitasking, better memory management, and many of the other features found in minicomputer operating systems. OS/2 1.0 was released in December 1987, and the second version, OS/2 1.10, released in October 1988, included a GUI called Presentation Manager. A severe shortage of memory microchips drove up the price of RAM memory from 1986 to 1989. In late 1988, a mere 1 megabyte of RAM cost about $900. OS/2 required substantially more memory than DOS, and the high costs of memory inhibited the widespread adoption of OS/2.

Even while working on OS/2 and Presentation Manager, Microsoft persisted in its own Windows efforts. Version 3.0, released in 1990, was an astounding commercial success, prompting pundits to argue that Microsoft took three tries to get its products right. Two factors contributed to the success of Windows: the memory shortage had ended, so more users found it easier to buy the extra memory that Windows demanded, and programmers at companies that made software applications had already been forced by the Macintosh and OS/2's Presentation Manager to learn how to program GUI programs. This programming knowledge easily transferred to writing software for the more successful Microsoft Windows.

IBM was never able to regain any momentum for OS/2, though OS/2 had matured into a solid operating system. When Microsoft decided to continue its Windows development efforts to the detriment of its OS/2 development efforts, Microsoft and IBM decided to sever the close partnership that had been characteristic in the 1980s. By this time, IBM was in deep disarray, as it had lost control of the PC market to the clone makers, found that PCs were becoming the dominant market segment in the computer industry, and saw the mainframe market begin to contract. IBM actually began to lose money and lost an astounding $8.1 billion on $62.7

billion in revenue in 1993. That year, IBM brought in a new chief executive officer from outside the company, Louis V. Gerstner Jr. (1942–), who managed to financially turn the company around through layoffs and refocusing the business on providing services. IBM remained the largest computer company but never dominated the industry as it once had. In contrast to IBM's size, Microsoft passed $1 billion a year in revenue in 1990.

By the late 1970s, people had seen television commercials for PCs. Apple II adopted the slogan, "Everyone should have a friend like Apple," and IBM used the Tramp (an imitation of Charlie Chaplin's character) confronting an IBM and quickly "keeping up with modern times" in 1981. They had also seen advertisements for games. Television, as a mass medium, had been used because PCs had become consumer products. By the early 1990s, the Windows product had become so popular that Microsoft started televising advertisements, which was unusual for an operating system.

Windows 3.0 was not really a new operating system, just a user interface program that ran on top of DOS. Microsoft created a new operating system, Windows NT, that contained the multitasking features, security features, and memory management that had made OS/2, UNIX, and other minicomputer operating systems so useful. Windows NT 3.1 came out in 1993, but it was not particularly successful until Windows NT 4.0 came out in 1996. Microsoft now had two Windows operating system lines, one for business users and servers and one for home consumers. With Windows 95, where Microsoft chose to change from version numbers based on release numbers to those based on years, Microsoft updated the consumer version of Windows. Windows 95 was an important product because the ease of use and aesthetic appeal promised by the GUI paradigm—and successfully achieved by Apple over a decade earlier—had finally been achieved by Microsoft.

Microsoft regularly produced new versions of its operating systems, adding features that demanded ever larger amounts of processor power, RAM, and disk drive space with each release. These increasing demands promoted the sales of ever more powerful PCs, making PCs effectively obsolete within only a few years of manufacture. The PC market in the 1990s and into the 2000s was dominated by what became known as the "Wintel alliance," a combination of the words *Windows* and *Intel*. With Windows XP, released in 2001, Microsoft finally managed to merge its consumer and business operating systems into a single release, after several earlier failed attempts.

In 1983, Microsoft released Microsoft Word, its word processing software application, principally written by a veteran from PARC, Charles Simonyi (1948–). At that time, products like MicroPro's WordStar and WordPerfect dominated the word processing market. Microsoft released a version of Word for the Macintosh in 1984 and came to dominate that market segment on the Macintosh, but Word did not threaten the success of other word processing applications on IBM PCs and PC clones until Windows 3.0 gave Microsoft developers a jump on the competition.

In the 1990s, Microsoft utilized its position as sole supplier of operating systems to PCs to compete against software application companies. Leading software products such as Lotus 1-2-3, Harvard Graphics, WordPerfect, and dBase began to lose market share after Windows 3.0 changed the PC market direction from the command-line DOS to Windows and consolidated three types of software packages (word processing, spreadsheets, and presentations) into a single package, Microsoft Office, released 1990. Lotus 1-2-3 3.0 and WordPerfect 5.1 were the best-selling software packages in electronic spreadsheets and word processors, respectively, in 1991. Only three years earlier, Lotus had earned more in gross revenue than all of Microsoft, but by 1991, they were roughly on par. By the year 2000, some version of Windows was on over 90 percent of the PCs in the world, and in application software, Microsoft's Excel and Word programs had replaced most of the market share once enjoyed by Lotus 1-2-3 and WordPerfect. IBM bought Lotus Development in 1995, mostly focusing on the Lotus Notes collaboration software, and maintained a version of 1-2-3 for its customers until 2014. WordPerfect products were still sold in 2022 to a marginal market, and one of the authors of this book remained a devoted fan.

Microsoft battled repeated complaints and lawsuits that it unfairly used its dominance in the operating systems market segment to dominate other PC software market segments. These complaints were based on two assertions. First, Microsoft created undocumented system calls that allowed its own applications to take special advantage of the Windows operating system. It had also done this in DOS. A federal court Finding of Fact in 1999 noted that IBM was not granted rights for Windows 95 until fifteen minutes before Microsoft released the operating system. A second finding was that Microsoft set up special deals with PC manufacturers, such as Compaq, Dell, and Gateway, where Microsoft sold its operating systems at a steep discount if the computer manufacturers only sold the Microsoft applications software at the same time.

These original equipment manufacturer (OEM) deals encouraged consumers to turn from buying their business applications software from retail stores to buying them from the computer manufacturer. The market for retail stores offering computer software collapsed, and those stores mostly disappeared during the 1990s. Egghead Software, for example, had begun in 1984 in Bellevue, Washington, and had over 200 stores at its height in the mid-1990s, but it had gone bankrupt by 2001. The federal government twice sued Microsoft for antitrust violations on its software distribution and pricing practices and both times found against Microsoft, but no effective legal counteraction was ever taken. Microsoft's practices made Bill Gates and his company widely disliked, yet respected, by many in the software development industry and computer industry, a disdain and respect once reserved for IBM.

Microsoft also aggressively entered any market that it thought might overshadow its dominance of PC software by possibly making PCs less important, launching a version of Windows for personal digital assistants, Windows CE, in the late 1990s, and a game console system called the Xbox in 2001. Windows CE failed, though Xbox became an important player in the game console industry. A Windows phone operating system also failed miserably.

Having developed Hodgkin's disease, Paul Allen left active participation in Microsoft in 1982. Until he died, Allen was still one of the richest men in the world and had sponsored and invested in many endeavors, including major sports teams; the Experience Music Project and Science Fiction Museum and Hall of Fame in 2004 (rebranded the Museum of Popular Culture (MoPOP) in 2016) in Seattle, devoted to his guitar idol, Jimi Hendrix, and his love of science fiction; and SpaceShipOne, the first commercially funded piloted vehicle to reach space.

Gates resigned as chief executive officer of Microsoft in 2000, naming himself chief software architect so that he could be more involved in the technical direction of the company and less distracted by its day-to-day management. Gates was the richest man in the world, worth more than $80 billion, though he had placed a substantial part of his fortune in a philanthropic trust. He stepped down as chairman in 2014.

In the early 2000s, Microsoft announced that it estimated that there were 600 million Windows PCs around the world, and it expected that number to pass 1 billion in just six more years. Microsoft revenues continued to set records. By 2004, Microsoft employed over 50,000 people and had a total annual revenue of over $35 billion, of which over $26 billion was gross profit. Microsoft's practice of not usually paying stock dividends meant that it had accumulated a cash reserve of $56 billion and zero debt. Microsoft employed 166,000 people worldwide by 2020 and had annual revenue of $68 billion; the total company was worth an astounding $1.5 trillion. A major driver of continued growth was cloud services. The little company that Gates and Allen had founded in 1975 had grown to become one of the most profitable on the planet and part of the Dow Jones 30 Industrials, the world's most commonly quoted stock indicator.

# SIX

# Connections: Networking Computers Together

## THE COLD WAR

The rocket engineers of the Soviet Union embarrassed the United States in 1957 by launching Sputnik 1, the first artificial satellite. This event provoked a strong political and cultural reaction in the United States; funding for education, especially science and engineering, increased, and federal funds for research and development in science and technology also rose. A space race rapidly emerged. As a struggle of competing ideologies, the Cold War conflict between the superpowers depended as much on prestige as military power, and the United States wanted to regain its prestige as the preeminent scientific and technological power on the planet.

The Advanced Research Projects Agency (ARPA) was formed in 1958 in response to Sputnik and the emerging space race. Because ARPA was an agency of the Pentagon, its researchers were given a generous mandate to develop innovative technologies. Though ARPA scientists and engineers did conduct their own research, much of the effort came through funding research at universities and private corporations.

In 1962, a psychologist from the Lincoln Laboratory at the Massachusetts Institute of Technology (MIT), J. C. R. Licklider, joined ARPA to take charge of the Information Processing Techniques Office (IPTO). Licklider's intense interest in cybernetics and "man-computer symbiosis" was driven by his belief that computers could significantly enhance the ability of humans to think and solve problems. Licklider (or "Lick," as he was called) created a social network of like-minded scientists and engineers and wrote a famous 1963 memorandum to these friends and colleagues, called "Memorandum for Members and Affiliates of the Intergalactic Computer Network," in which he described some of his ideas for time-sharing and computer networking. His IPTO funded research efforts in time-sharing, graphics, artificial intelligence (AI), and communications, laying the conceptual and technical groundwork for computer networking.

## TELEPHONES

Networking already existed in the form of telegraphs and telephones. Samuel F. B. Morse (1791–1872) invented the telegraph in 1844, allowing communications over a copper wire via electrical impulses that operators sent as dots and dashes. Alexander Graham Bell (1847–1922) invented the telephone in 1877, using an analog electrical signal to send voice transmissions over copper wires. Teletype systems were first patented in 1904 and allowed an automatic typewriter to receive telegraph signals and print out the message without a human operator.

In the 1950s, the U.S. military wanted its new Semi-Automatic Ground Environment (SAGE) computers to communicate with remote terminals, so engineers developed a teletype to send an analog electrical signal to a distant computer. In 1958, researchers at Bell Telephone Laboratories took the next step and invented the modem, which stood for modulator-demodulator. Modems converted digital data from a computer to an analog signal to be transmitted across phone lines and then converted that signal back into digital bits for the receiving computer to understand. In 1962, the Bell 103, an early commercial modem, was introduced to the market by American Telephone and Telegraph (AT&T), the parent company of Bell Labs, running at 300 baud, which transmitted 300 bits per second. Modem speed steadily increased, eventually reaching 56K in the mid-1990s.

Each computer manufacturer tended to define its own character set for both letters and numbers, even changing them from model to model, forcing programmers to convert data when transferring their files from one computer to another. The American National Standards Institute (ANSI) defined the American Standard Code for Information Interchange (ASCII) in 1963. This meant that the binary sequence for the letter *A* would be the same on all computers. While the rest of the industry turned to ASCII, especially when networking and PCs became more common in the 1970s, IBM maintained its own standard, Extended Binary-Coded Decimal Interchange Code (EBCDIC), for decades.

## PACKET SWITCHING

In the early 1960s, the Polish-born electrical engineer Paul Baran (1926–2011), who worked for the RAND Corporation, a think tank funded by the American military, faced a problem. Simulations of an attack with nuclear weapons by the Soviet Union showed that even minor damage to the long-distance phone system maintained by the telephone monopoly AT&T would cripple national communications. The telephone system that had developed during the twentieth century was based on analog transmissions over lines connected to switches. When a person made a long-distance telephone call, an actual electrical circuit was created via numerous switches in a scheme called *circuit switching.*

Baran had considerable experience with computers, including working on the original UNIVAC, and appreciated the value of digital electronics over analog electronics. Starting in 1959, Baran devised a scheme of breaking signals into blocks of data to be reassembled after reaching their destination. These blocks of data traveled through a *distributed network* where each *node*, or communication point, could independently decide which path the block of information took to the next node. This allowed data to automatically flow around potential blockages in the network and be reassembled into a complete message at the destination. Baran called his scheme "hot potato" routing, because each network node tossed the message to another node rather than hold on to it.

The Pentagon and AT&T were not interested in Baran's scheme of distributed communications because it required completely revamping the technology of the national telephone system. A British team under the direction of Donald Davies (1924–2000), at the British National Physical Laboratory (NPL), also independently developed a similar scheme to Baran's, which they called *packet switching.* Davies and his team went further than Baran and actually implemented their ideas, and by 1970, they had a local area network running at the NPL that used packet switching.

## ARPANET

In 1966, Robert "Bob" Taylor, then head of the IPTO, noted that in his terminal room at the Pentagon, he needed three different computer terminals to connect to three different machines in different locations around the nation. Taylor also recognized that universities working with the IPTO needed more computing resources. Instead of the government buying large and expensive computers for each university, why not share the larger computers? Taylor revitalized Licklider's ideas, secured $1 million in funding, and hired Larry Roberts (1937–2018), a twenty-nine-year-old computer scientist, to direct the creation of ARPANET.

In 1965, while working at MIT's Lincoln Laboratory, Roberts had supervised an ARPA-funded pilot project to have two computers communicate over a long distance. Two computers, one in Boston and the other in Santa Monica, California, sent messages to each other over a set of leased Western Union telephone lines. The connection ran slowly and unreliably, but it offered a direction for the future. ARPANET was the next logical step. Roberts drew on the work of Baran and Davies to create a packet-switched networking scheme. While Baran was interested in a communications system that could continue to function during a nuclear war, ARPANET was purely a research tool, not a command and control system, as was sometimes reported in contemporary media accounts.

Universities were reluctant to share their precious computing resources and concerned about the processing load of a network on their systems. Wesley Clark (1927–2016), the computer lab director at Washington University in St. Louis, proposed an Interface Message Processor (IMP), a separate smaller computer for each main computer on the network that would handle the network communication. This also allowed for a consistent technology as an interface to very different machines.

A small consulting firm in Cambridge, Massachusetts, Bolt Beranek and Newman (BBN), got the contract to construct the needed IMPs in December 1968. It decided that the IMP would only handle the routing, not the transmitted data content. As an analogy, the IMP only looked at the addresses on the envelope, not at the letter inside. Faculty and graduate students at the host universities created host-to-host protocols and software to enable the computers to understand each other. Because the machines did not know how to talk to each other as peers, the researchers wrote programs that fooled the computers into thinking they were talking to preexisting "dumb" terminals.

ARPANET began with the installation of the first 900-pound IMP, which cost about $100,000 to build, in the fall of 1969 at the University of California, Los Angeles (UCLA), followed by three more nodes at the Stanford Research Institute (SRI); the University of California, Santa Barbara; and the University of Utah by December. Fifty kilobit per second (kbps) communication lines connected each node to the others. The first message, transmitted on October 29, between UCLA and SRI's Scientific

Data Systems (SDS) computers was "L-O-G," the first three letters of LOGIN, then the system crashed. Initial bugs were overcome within an hour, and a connection was made. After that, ARPANET added an extra node every month in 1970. BBN continued to run ARPANET for the government, keeping the network running through round-the-clock monitoring at its network operations center.

With a network in place, ARPANET scientists and engineers turned to using the network to get useful work done. Transferring files and remote login were obvious and useful applications. In 1971, file transfer protocol (FTP) was developed. The protocol originally required a user to authenticate themselves with a username and password, but a system of using anonymous FTP later allowed any user to download files that had been made available for everyone. Remote login was achieved through a variety of programs, although telnet, developed in 1971, eventually became the standard.

Also in 1971, Ray Tomlinson (1941–2016), an engineer at BBN supporting ARPANET, found himself working on a program called CPYNET (for "copynet"), which was designed to transfer files between computers. He realized that CPYNET could be combined with SNDMSG (for "send message")—a program designed to send messages to a user on the same computer—and send messages from one computer to another. Tomlinson did so, and email (electronic mail) was born. Tomlinson also developed the address format user@computer that used the @ symbol and later became ubiquitous. Electronic mail became what later pundits would call the *killer app* of ARPANET, its most useful feature and its most commonly used application.

From the beginning of networking, programs had been designed to run across the network. As time went by, many of these programs used the same design structure, which became known as client-server systems by the early 1990s. A server program provided some service, such as a file, email, or connection to a printer, while a client program communicated with the user and the server program so that the user could use the service.

Roberts succeeded Taylor as head of the IPTO, and in 1972, he arranged for a large live demonstration of ARPANET at the International Conference on Computer Communications in Washington, DC. None of the work on ARPANET was classified, and the technical advances from the project were freely shared. The vision of what was possible with networking rapidly caught the imagination of scientists and engineers in the rest of the computer field. IBM announced its own Systems Network Architecture

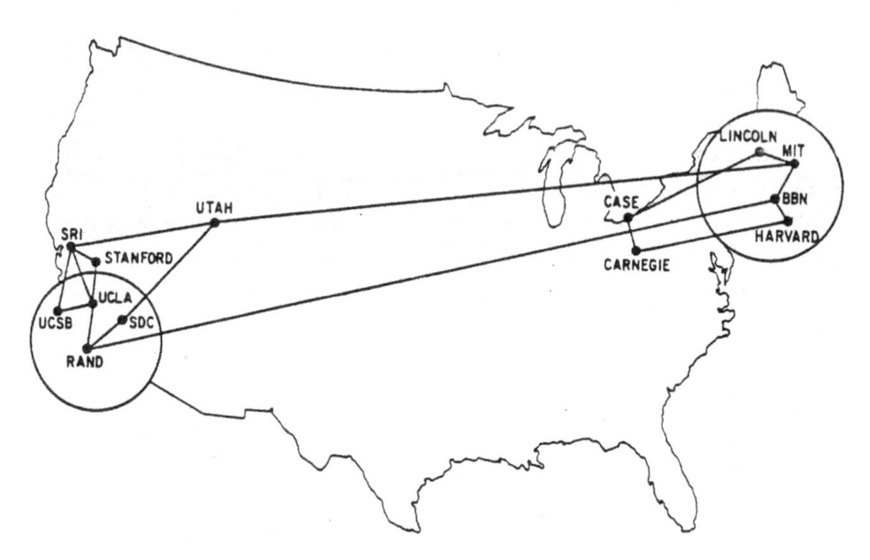


ARPANET in December 1970. (Image Courtesy of the Charles Babbage Institute Archives, University of Minnesota Libraries, Minneapolis)

(SNA) in 1974, which grew more complex and capable with each passing year. Digital Equipment Corporation (DEC) released its DECnet in 1975, implementing its Digital Network Architecture (DNA). Other large computer manufacturers also created their own proprietary networking schemes.

## THE BEGINNING OF WIRELESS AND UBIQUITOUS NETWORKING

ARPA also funded the effort by Norman Abramson (1932–2020) of the University of Hawaii to build AlohaNet in 1970. In addition to being one of the earliest packet-switching networks, AlohaNet broke new ground as the first wireless network. AlohaNet transmitted radio signals between terminals, the computer taking over a role similar to ham radio enthusiasts sharing the same frequency. One of the first technical hurdles to such a scheme was, how does the network program on a terminal know when it can send a radio signal? If two terminals sent a signal at the same time, the signals would interfere with each other, becoming garbled, and neither would be received by other terminals. The conventional answer was time-division multiple access (TDMA), where terminals coordinated their activity and only transmitted during their allocated time. For instance, perhaps each terminal would each get a fraction of a second, and no two terminals could use the same fraction. The problem with this scheme was how to actually divide up the time slices and account for some terminals being used while others were offline. TDMA tended to become more difficult to maintain as more terminals were added to the conversation.

AlohaNet had so many terminals that TDMA was impractical, so a new scheme was developed: carrier-sense multiple access with collision detection (CDMA/CD). Under this scheme, any terminal could transmit whenever it wanted, but then it listened to see whether its transmission was garbled by another transmission. If the message went through, everything was fine, and the bandwidth was now free for any other terminal to use; if the signal became garbled, the sending terminal recognized that it had failed and waited for a random amount of time before trying to send the same message again. This scheme, seemingly chaotic, worked well in practice as long as there were not too many terminals and as long as traffic was low enough so that there were not too many collisions. The scheme also allowed terminals to readily be added to and removed from the network without needing in any way to inform the other terminals about their existence.

Robert Metcalfe (1946–), a researcher at the innovative Xerox Palo Alto Research Center (PARC), visited Hawaii in 1972 and studied AlohaNet for his doctoral dissertation. Returning to Xerox, Metcalfe then developed Ethernet using the CDMA/CD scheme running over local wire networks. Metcalfe left Xerox to cofound 3Com in 1979, a company that successfully made Ethernet the dominant networking standard on the hardware level in the 1980s and 1990s.

## TCP/IP AND RFCS

ARPANET originally used a set of technical communications rules called the network control protocol (NCP). NCP assumed that every main computer on the ARPANET had identical IMP computers in front of them to take care of the networking. All the IMP machines were built by the same people using the same designs, minimizing the risk of incompatibilities. This worked well, but NCP was not the only networking protocol available. Other companies developing their own networking schemes also developed their own set of proprietary protocols. Engineers at both BBN and the Xerox PARC wanted to create a new set of network protocols that would easily enable different networks, each running their own set of unique protocols (such as NCP or SNA), to communicate with each other. This idea, called *internetworking*, would allow the creation of a network of networks.

Vint Cerf (1943–) became known as one of the "fathers of the Internet." As a graduate student, he worked on the first IMP at UCLA and served as a member for the first Network Working Group that designed the software for the ARPANET. Bob Kahn (1938–) and Cerf first proposed Transmission Control Protocol (TCP) in 1974 to solve the problem of internetworking, and Cerf drove the further development of protocols in the 1970s. The internet working protocol eventually split into two parts: TCP and Internet Protocol (IP). TCP/IP was an open protocol, publicly available to everyone, with no restrictive patents or royalty fees attached to it, and ARPANET switched to using TCP/IP in the late 1970s.

The philosophy behind Metcalfe's Ethernet heavily influenced TCP/IP. The NCP scheme had little error correction because it expected the IMP machines to communicate reliably. TCP/IP could not make this assumption and included the ability to verify that each packet had been transmitted correctly. For TCP/IP to work correctly, each machine must have a unique IP address, which came in the form of four groups of numbers, each within the range 0–255, for instance, 192.168.54.213. TCP/IP also made it simple to add and remove computers to the network, just as Ethernet could. In July 1977, an experiment with a TCP system successfully transmitted packets via the three types of physical networks that made up ARPANET: radio, satellite, and ground connections. The packets began in a moving van in San Francisco, were transmitted via radio, crossed the Atlantic Ocean to Norway via satellite, bounced to London, and then returned to the University of Southern California, a total of 94,000 miles in transit. This proof of concept became the norm as the ARPANET matured.

The original team working on ARPANET was called the Network Working Group (NWG), which evolved into the Internet Engineering Task Force (IETF) and the Internet Engineering Steering Group (IESG). These groups used the unique process of Requests for Comments (RFC) to facilitate and document their decisions. The first RFC was published in 1969.

By 1989, with some 30,000 hosts connected to the internet, 1,000 RFCs had been issued. Ten years later, millions of hosts used the internet, and over 3,000 RFCs had been reached. The RFC process created a foundation for sustaining the open architecture of the ARPANET/internet, where multiple layers of protocols provided different services. Jon Postel (1943–1998), a computer scientist with long hair and a long beard, edited the RFCs for almost thirty years before his death in 1998, a labor of love that provided a consistency to the evolution of the internet. The actual work of the IETF continues to still be performed in working groups, and anyone may join a working group and contribute their observations and work to the group, which can result in a new RFC.

Postel worked on the internet for the rest of his life. One employer was the Information Sciences Institute (ISI), founded by the University of Southern California in 1972 as a nonprofit research center whose funding came exclusively from ARPA. As TCP/IP became the networking foundation of ARPANET, numbers had to be assigned for the many technical applications. RFC 790, "Assigned Numbers," issued in September 1981, explained that "current information can be obtained from Jon Postel. The assignment of numbers is also handled by Jon. If you are developing a protocol or application that will require the use of a link, socket, port, protocol, or network number please contact Jon to receive a number assignment." This delightful informality in such documents existed because the technical community supporting ARPANET and TCP/IP was still small enough to be informal. A couple of years later, a colleague of Postel's at ISI, Joyce K. Reynolds (1952–2015), took over the number assignment tasks, though Postel remained responsible. In 1987, the Pentagon moved responsibilities away from Postel and Reynolds and gave them to SRI.

## INTERNET

In the 1960s, after introducing the modem, AT&T began to develop the technology for direct digital transmission of data, avoiding the need for modems and the inefficiency that came from converting to and from analog. A lawsuit led to the Carterphone Decision in 1968, which allowed non-AT&T data communications equipment to be attached to AT&T phone lines, prompting other companies to develop non-AT&T modem and data communications equipment. In the 1970s, leased lines providing the digital transmission of data became available, including X.25 lines based on packet-switching technology. The availability of these digital lines laid the foundation for the further spread of wide area networks (WANs).

ARPANET was not the only large network, only the first that paved the way and eventually became the internet. International Business Machines (IBM) funded the founding of BITNET in 1984 as a way for large universities with IBM mainframes to network together. Within five years, almost 500 organizations had 3,000 nodes connected to BITNET, yet only a few years later, the network had disappeared into the growing internet. The Listserv program first appeared on BITNET to manage email lists, allowing people to set up, in effect, private discussion groups. These email lists could either be moderated or unmoderated. Unmoderated lists allowed anyone who wanted to join and send messages; moderated lists set up a person or persons as moderators, who controlled who could join the list and checked every email that went through the list before passing them on to the general membership of the list. Moderated lists became more popular because they prevented a flood of superfluous emails from dominating the list and driving away members.

In 1981, the National Science Foundation (NSF) created the Computer Science Network (CSNET) to provide universities that did not have access to ARPANET with their own network. In 1986, the NSF sponsored the NSFNET "backbone" to connect five supercomputing centers together at 56 kilobits per second. The backbone also connected ARPANET and CSNET. The idea of the internet, a network of networks, became firmly entrenched. The open technical architecture of the internet allowed numerous innovations to easily be grafted onto the whole, and proprietary networking protocols were abandoned in the 1990s as everyone moved to using TCP/IP. The NSFNET saw its first commercial internet service provider (ISP) out of Brookline, Massachusetts, The World, in 1989. By that time, parts of the backbone ran at just 1.5 megabits per second (called *T-1*).

As mentioned before, TCP/IP only recognized different computers' hosts by their unique IP numbers, such as 192.168.34.2. People are not very good at remembering arbitrary numbers, so a system of giving computers names quickly evolved. It had begun simply enough with a file at SRI called HOSTS.TXT during the early days of ARPANET. Each computer using TCP/IP had a system file on it called "hosts" that contained entries matching the known names of other computers to their IP addresses. These files were each maintained individually, and the increasing number of computers connected to the ARPANET/internet created confusion. In 1983, a domain name system (DNS) was created, where DNS servers kept master lists matching computer names to IP addresses for a particular zone. A hierarchical naming system was also created, with computer names being attached to domain names and ending with the type of domain. Six extensions were created:

.com: commercial
.edu: educational
.net: network
.gov: government
.mil: military
.org: organization

While ARPANET was very much an American creation and mostly existed only in the United States, the engineers and scientists expanding the network envisioned an expanded worldwide presence. An RFC in 1984 listed other top-level domains in the form of two-letter codes to represent countries, such as ".fr" for France and ".jp" for Japan. At that time, there were only a thousand hosts on the ARPANET. When ARPANET was physically dismantled in 1990 and replaced by newer technology, the internet was thriving at universities and technology-oriented companies. In 1991, the federal government lifted the restriction on the use of the internet for commercial use. The NSF backbone was later dismantled in 1995 when the NSF realized that commercial entities could keep the internet running and growing on their own. The NSF backbone had cost only $30 million in federal money during its nine-year life, with donations and help from IBM and MCI (a telecommunications company). What began with four nodes in 1969 as a creation of the Cold War became a worldwide network of networks that formed a single whole. An estimated 120 million computers were connected to the internet in every country of the world in early 2001. That number was over 10 billion in 2021. As a global computer network interconnecting other computer networks, the internet provided a means of communication unprecedented in human history.

## BULLETIN BOARD SYSTEMS AND DIAL-UP PROVIDERS

In January 1978, when a severe snowstorm shut down the city of Chicago, two friends, Ward Christensen (1945–) and Randy Suess (1945–2019), decided to develop a system to exchange messages. Christensen wrote the software, and Suess put together the hardware, based on a homemade computer using a S-100 bus and hand-soldered connections running the CP/M operating system. They finished their effort in a month and called their system the Computer Bulletin Board Systems (CBBS), which allowed people to call in, post messages, and read messages. Modems at the time were rare, but the subsequent development of cheaper modems allowed computer hobbyists to set up their own Bulletin Board Systems (BBSs) and dial into other BBSs. Later enhancements allowed users to upload and download files, enter chat areas, and play games. Hundreds of thousands of BBSs eventually came and went, serving as a popular communications mechanism in the 1980s and early 1990s. A separate network connecting BBSs even emerged in the mid-1980s, called FidoNet, that allowed users to exchange email and discussion messages. At its height in 1995, FidoNet connected some 50,000 BBS nodes to each other. The coming of the public internet in the 1990s doomed the BBS as a technology, though the social and special interest communities that had grown up around various BBSs transferred their communities to the internet.

In 1969, CompuServe began as a time-sharing service in Columbus, Ohio. A decade later, in 1979, the company expanded to offer email and simple services to home users of PCs. In 1980, CompuServe offered the first real-time chat service with a program called CB Simulator that allowed users to simultaneously type in messages and have the results appear on each other's screens. From this humble beginning, what later became instant messaging was born. In the 1980s, CompuServe built its own countrywide network, which customers could use by dialing in with a modem to connect to large banks of modems that CompuServe maintained. CompuServe also offered the use of its network to corporations as a WAN and expanded into Japan and Europe. CompuServe continually expanded the offerings that its customers paid to access, such as discussion groups, content from established national newspapers and magazines, stock quotes, and even a stock trading service.

Sears and IBM created Prodigy, their own online service provider, in the 1980s, and it soon had over a million subscribers. In 1985, Steve Case (1958–), a computer enthusiast with a taste for business and filled with entrepreneurial zeal, joined Control Video Corporation (CVC), a company oriented toward the Atari 2600. Jim Kimsey (1939–2016) refashioned CVC into Quantum Computer Services, a BBS, and PlayNET, which allowed for multiplayer games, for users of Commodore 64 PCs. When Case wanted to expand and compete with the other online services, like CompuServe and Prodigy, he pushed his company to become America Online (AOL) in 1989 and took over as CEO in 1991.

After the federal government lifted the restriction on the use of the internet for commercial use in 1991, numerous internet service providers (ISPs) sprang up immediately to offer access to the internet for a monthly fee. The internet included a million hosts by 1992. CompuServe, AOL, and Prodigy began to provide access to the internet to their customers, thus transforming these companies into instant ISPs. CompuServe became the largest ISP in Europe. Fueled by an aggressive marketing campaign, which included flooding the nation's mail with sign-up disks, AOL grew quickly, reaching one million subscribers in August 1994, passing two million subscribers in February 1995, and peaking at twenty-five million subscribers in 2000. Over one million of those subscribers were in Germany, and AOL had over five million subscribers outside the United States in 2001. AOL grew so large that it purchased CompuServe in 1997. Prodigy failed to successfully make the transition to being an ISP and faded away. AOL merged with Time Warner in 2001, and Case retired from the chairman position of AOL Time Warner in 2003. AOL became part of telecommunications giant Verizon in 2015.

## LOCAL AREA NETWORKS

The APRANET and its successor, the internet, are examples of WANs, where computers communicated across the street, across the nation, and even around the world. In the 1970s, research at the Xerox PARC, which had led to many innovations, also led to the creation of local area networks (LANs). A LAN was usually defined as a network for a room or a building. Ethernet provided one of the early standards for LAN computing, though other LAN network technologies, such as Token Ring and ARCNET, also appeared.

In the early 1980s, with the easy new availability of a large number of PCs, various companies developed network operating systems (NOS), mainly to provide an easy way for users to share files and printers. Later, LAN-based applications based on the NOS became available. The most successful of these NOS came from Novell. The company, founded in 1979 as Novell Data Systems, originally made computer hardware, but after the company was sold in 1983, the new president of the company, Raymond J. Noorda (1924–2006), turned it toward concentrating on software. That same year, the first version of NetWare came out. The package allowed companies to easily network computers with various services, such as email.

Novell ruled the LAN NOS market for the next decade, achieving almost a 70 percent share and adding ever more sophisticated features to each version of NetWare. Novell created its own networking protocols, called IPX (Internet Packet eXchange) and SPX (Sequenced Packet eXchange), drawing on open networking standards that Xerox had published. Eventually, in the 1990s, as the internet became ever more pervasive, Novell also turned to supporting TCP/IP as a basic protocol in NetWare. The dominance of NetWare rapidly declined in the late 1990s when Microsoft provided networking as a basic part of its Windows operating systems. Novell and Microsoft talked for years about a possible merger, but Microsoft eventually decided to go its own way.

## USENET

In 1979, graduate students at Duke University and the University of North Carolina wrote some simple programs to exchange messages between UNIX-based computers. This collection of programs, called *news*, allowed users to post messages to a newsgroup and read messages that other users had posted to that same newsgroup. The news program collected all the postings and then regularly exchanged them with other news programs via homemade 300-baud modems. The students brought their project to a 1980 USENIX conference. At that time, ARPANET was only available to universities and research organizations that had defense-related contracts, so most universities were excluded from the network. Because so many universities had UNIX machines, USENIX conferences allowed users to meet and exchange programs and enhancements for the UNIX operating system itself. The students proposed that a "poor man's ARPANET" be created, called USENET, based on distributing the news program and using modems to dial up other UNIX sites. To join USENET, one just had to find the owner of a USENET site who would allow you to download a daily news feed.

USENET grew quickly, reaching 150 sites in 1981, 1,300 sites by 1985, and 11,000 sites in 1988. A protocol was eventually developed in the early 1980s, the Network News Transfer Protocol (NNTP), so that news reader clients could connect to news servers. ARPANET sites even joined USENET because they liked the USENET newsgroups. An entire culture and community grew up around USENET, where people posted technical questions on many aspects of programming or computing and received answers within a day from other generous users. USENET news discussion groups originally concentrated on technical issues, then it expanded into other areas of interest. Anyone who wanted to could create a new newsgroup, though if it did not attract any postings, the newsgroup eventually expired. Programmers developed a way to encode binary pictures into ASCII, which could then be decoded at the other end, and picture newsgroups, including an enormous number of pornographic pictures, became a major part of the daily USENET news feed.

As part of the culture of USENET, a social standard of net etiquette evolved, eventually partially codified in 1995 in RFC 1855, "Netiquette Guidelines." One such rule is that words in all-capital letters are the equivalent of shouting. Excessive and personal criticism of another person in a newsgroup came to be called "flaming," and "flame wars" sometimes erupted, the equivalent of an online shouting match, with reasoned discourse abandoned in favor of name-calling. A set of acronyms also evolved, such as IMO for "in my opinion" and LOL for "laugh(ing) out loud," as well as some symbols for expressing emotions (emoticons), such as :-) for a smile and :-(for a frown. Communicating by graphic characters took hold in computing. By 1997, emojis (from the Japanese *e* (picture) and *moji* (character)) could be sent on Japanese phones, and by 2020, almost 3,000 emojis were available as part of the Unicode character set.

The number of USENET messages exploded in the 1990s, especially after AOL created a method for its millions of subscribers to access USENET, but the usefulness of USENET declined in proportion to the number of people using it. The flooding of newsgroups with advertising messages also drove away many people, who found refuge in email list servers, interactive websites, and private chat rooms. A major component of the success of USENET came from the fact that most people did not have access to the internet; when that access became more common, USENET no longer offered any serious advantages. The etiquette standards created for USENET have continued, being applied to BBS chat rooms, web-based chat rooms, and informal email. As social media sites emerged on the internet in the early 2000s, users and administrators found themselves struggling with the same user etiquette issues.

## GOPHER

As the internet grew ever larger in the 1980s, various schemes were advanced to make finding information content on the internet easier. The problem of finding content even existed on individual university campuses, and in the 1980s, various efforts were made to solve the problem on a smaller scale through Campus-Wide Information Systems (CWIS). Cornell University created CUinfo, Iowa State created CyNet, and Princeton created PNN, all early efforts to organize information.

Programmers at the University of Minnesota released the Gopher system in April 1991 to solve this problem. Gopher consisted of Gopher servers to hold documents and Gopher clients to access the documents. The system interface was entirely based on simple ASCII text and used a hierarchy of menus to access documents. The creators of Gopher thought of their creation as a way to create a massive online library. Anyone could download the server software, organize their content into menus and submenus, and set up their own Gopher server. Pictures and other multimedia files could be found and downloaded through Gopher but not displayed within the client. Gopher's virtues included a lean interface and a transmission protocol that did not strain the limited network bandwidth that most systems suffered from.

Gopher quickly grew in popularity as people on the internet downloaded the free software and set up their own Gopher servers. Gopher software was rapidly ported to different computer models and operating systems. Even the Clinton-Gore administration in the White House, enthusiastic to promote what they called the "information highway," announced its own Gopher server in 1993. Gopher was the first application on the internet that was easy to use and did not require learning a series of esoteric commands. Users enjoyed "browsing," going up and down menus to find what gems of text a new Gopher server might offer.

The problem of how to find content reemerged. All these different Gopher servers were not connected in any way, though the "Mother Gopher" server at the University of Minnesota had some links in its menus to other Gopher servers. Late in 1992, a pair of programmers at the University of Nevada, Reno, introduced Veronica. The name came from the *Archie* comic book series, but to make the word into an acronym, they came up with the Very Easy Rodent-Oriented Netwide Index to Computerized Archives (VERONICA). Veronica searched the internet for Gopher files, indexed them, and allowed users to search those indexes through a simple command-line interface. An alternate indexing program from the University of Utah was called Jughead, again drawing on the *Archie* comic books.

The number of known Gopher servers grew from 258 in November 1992 to over 2,000 in July 1993 and almost 7,000 in April 1994. In the spring of 1993, the administration of the University of Minnesota, having financially supported the creation of Gopher, decided to recover some of its costs by introducing licensing. The license kept Gopher software free for individual use but charged a fee for commercial users based on the size of their company. Considerable confusion surrounded this effort, and it sent a chill over the expansion of the protocol. Meanwhile, another protocol, based on hypertext documents, had been introduced to solve the same problem as Gopher, and Gopher faded away.

## WORLD WIDE WEB

Tim Berners-Lee was born in London to mathematician parents who had both worked as programmers on one of the earliest computers, the Manchester Mark I, at Manchester University. He graduated with honors and a bachelor's degree in physics from Oxford University in 1976. In 1980, he went to work at the Conseil Européen pour la Recherche Nucléaire (CERN), a nuclear research facility on the French-Swiss border, as a software developer.

The physics community at CERN used computers extensively, with data and documents scattered across a variety of different computer models, often created by different manufacturers. Communication between the different computer systems was difficult. A lifelong ambition to make computers easier to use encouraged Berners-Lee to create a system to allow easy access to information. He built his system on two existing technologies: computer networking and hypertext. Hypertext was developed in the 1960s by a development team at Stanford Research Institute (SRI) led by the computer scientist Douglas C. Engelbart and others and was based on the idea that documents should have hyperlinks in them that connect to other relevant documents, allowing a user to navigate nonsequentially through content. The actual word *hypertext* was coined by Ted Nelson (1937–) in the mid-1960s.

Using a new NeXT PC, with its powerful state-of-the-art object-oriented programming tools, Berners-Lee created a system that delivered hypertext over a computer network using the hypertext transfer protocol (HTTP). He simplified the technology of hypertext to create a display language that he called hypertext markup language (HTML). The final innovation was to create a method of uniquely identifying any particular document in the world. He used the term *universal resource identifier* (URI), which became *universal resource locator* (URL). The very first website was http://info.

cern.ch/hypertext/WWW/TheProject.html. In March 1991, Berners-Lee gave copies of his new World Wide Web (WWW) programs, a web server and text-based web browser, to his colleagues at CERN. By that time, internet connections at universities around the world were common, and the WWW caught on quickly as other people readily converted the necessary programs to different computer systems. The Stanford Linear Accelerator Center (SLAC) became the second website.

The WWW proved to be more powerful than Gopher in that hypertext systems are more flexible than hierarchical systems and more closely emulate how people think. Gopher had an initial advantage in that its documents were simple to create: they were just plain ASCII text files. Creating web page files using the Berners-Lee scheme required users to learn HTML and manually embed formatting commands in their pages. However, Berners-Lee also accommodated this, as he included Gopher as one of the protocols that web browsers could access. By using "gopher://" instead of "http://," he thus effectively incorporated Gopher in the emerging WWW.

A team of staff and students at the National Center for Supercomputing Applications (NCSA) at the University of Illinois at Urbana-Champaign released a graphical web browser called Mosaic in February 1993, making the WWW even easier to use. For a time, character-mode web browsers, like Lynx, were popular, but the increasing availability of bitmapped graphics monitors on PCs and workstations soon moved most users to the more colorful and user-friendly graphical browsers. The WWW made it easy to transfer text, pictures, and multimedia content from computer to computer. The creation of HTML authoring tools made it easier for users to create web pages without fully understanding HTML syntax or commands. This became more important as HTML underwent rapid evolution, adding new features and turning what had been relatively simple markup code into complex-looking programming code that supported tables, frames, style sheets, and JavaScript—a procedural programming language. Berners-Lee's original vision of the WWW included the ability for consumers to interactively modify the information that they received, though this proved technically difficult and has never been fully implemented.

The WWW became the technology that made the internet accessible to the masses, becoming so successful that the two terms became interchangeable in the minds of nontechnical users. Even technical users, who knew that the internet was the infrastructure and the WWW was only a single protocol among many protocols on the internet, often used the two terms interchangeably. Because of slow network speeds, graphics-intensive web pages could take a long time to load in the web browser, leading many to complain that WWW stood for "world wide wait."

A major key to the success of the WWW came from generosity on the part of CERN and Berners-Lee to not claim any financial royalties for the invention, unlike the remunerative-oriented efforts of the University of Minnesota with its Gopher technology. Berners-Lee moved to the Massachusetts Institute of Technology (MIT) in 1994, where he became director of the WWW Consortium. This organization, under the guidance of Berners-Lee, continued to coordinate the creation of new technical standards to enable the WWW to grow in ability and power. New programming standards for web pages were proposed and adopted, using the RFC system, making the WWW ever more versatile and complex. Berners-Lee became a hero in his native Britain and to the computer community at large. He won many awards and even had a role in the opening ceremonies of the 2012 London Summer Olympics.

An internet economy based on the WWW emerged in the mid-1990s, dramatically changing many categories of industries within a matter of only a few years. Members of the original Mosaic team, including Marc Andreessen (1971–), moved to Silicon Valley in California to found Netscape Communications in April 1994 with Jim Clark (1944–). Clark had already made a fortune from founding Silicon Graphics Inc. (SGI) in 1982, a high-end maker of UNIX computers and software used in 3-D graphics-intensive processing. Netscape brought out one the first commercial web browsers and also created JavaScript, rapidly developing the technology by adding new features; it became the dominant web browser.

Bill Gates at Microsoft recognized during a personal retreat in 1994 that a web browser had the potential to add features and grow so sophisticated that it could take over the role of an operating system. This threatened the foundation of Microsoft's success, and Gates reacted by turning his company from being focused on just PC software to an internet-centric vision. Before this time, Microsoft had concentrated on creating its own online service to compete with AOL and CompuServe, called Microsoft Network. Microsoft was so tardy in understanding the internet that its first internet site, a file repository for customer support, was not created until early 1993. Microsoft only happened to own its own domain name, Microsoft.com, because an enterprising employee had registered it during the course of writing a TCP/IP networking program.

As part of Gates's strategy, Microsoft released its own web browser, Internet Explorer (IE), in 1995, offering it for free. Early Microsoft browsers were not technically on par with Netscape, but after several years, IE became a more solid product. Microsoft also made strong efforts to integrate (or bundle) IE into its operating system. Doing so allowed it to leverage its monopoly in PC operating systems and force Netscape from the marketplace. Netscape was sold to AOL in 1998, mostly for the value of its high-traffic web portal, Netscape.com, rather than the declining market share of its browser. Microsoft's tactics also led to a famous antitrust lawsuit by the federal government, which dragged on from 1997 to 2004 and ended with minor sanctions on Microsoft.

Netscape's initial public stock offering in August 1995 had turned the small company into a concern valued at several billion dollars. This symbolized the emerging *dot-com* boom in technology stocks. Billions of dollars of investment poured into internet-based start-ups, based on the belief that the internet was the new telegraph or railroad and that those companies that established themselves first would be the ones that grew the largest. Many young computer technologists found themselves suddenly worth millions, or even billions, of dollars. In such an exuberant time, with speculation driving up stock prices around the world, some pundits even predicted that traditional rules of business had evolved and no longer applied where companies should focus on market share and not profit. One of the best examples of dot-com exuberance came in January 2001 when AOL completed its merger with the venerable Time Warner media company, a deal based entirely on AOL's high stock valuation, which quickly became a financial disaster after AOL's stock value crashed. Alas, in the end, a company must eventually turn a profit.

The dot-com boom ended in late 2000, a bursting of the stock market bubble, which caused an economic contraction and depression within the computer industry and contributed to an economic recession in the United States. It also lessened the percentage of young Americans interested in pursuing careers in computing, a situation that only turned around vigorously later that decade when computer science departments saw a renewed influx of students.

Amid the litter of self-destructing dot-coms, fleeing venture capitalists, and the shattered dreams of business plans, some dot-com companies did flourish. Amazon.com established itself as the premier online bookstore, fundamentally changing how book buying occurred, and eBay.com found a successful niche offering online auctions. PayPal provided a secure mechanism to make large and small payments on the web, and DoubleClick succeeded by providing software tools to obtain marketing data on consumers who used the WWW—and by also collecting those data itself. The end of the dot-com boom also dried up a lot of the money that had flowed into web-based advertising. After the dot-com crash, the WWW and internet continued to grow, but commerce on the web, dubbed *e-commerce*, grew at a slower rate dictated by more prudent business planning.

## WEB SEARCH ENGINES

Just as Gopher became really useful when Veronica and Jughead were created as search programs, the WWW became more useful as web-crawling programs were used to create web search engines. These programs prowled the internet, trying to divine the purpose of web pages by using the titles of the pages, keywords inside HTML meta tags embedded within the web page itself, and the frequency of uncommon words in the page to determine what the page was about. When users used a search engine—such as the early www.webcrawler.com or www.altavista.com—a database of results from relentless web-crawling software, sometimes called *spiders*, returned a list of suggested websites ranked by probable matches to the user's search words. Early search engines became notorious for at times returning the oddest results, but they were better results than having nothing.

The other approach to indexing the web was by hand, using humans to decide what a web page was really about and connecting that page to keywords in the search engine databases. A pair of PhD candidates in electrical engineering at Stanford University, David Filo (1966–) and Taiwanese-born Jerry Yang (1968–), created a website called Jerry's Guide to the WWW. This list of links grew into a large farm of web links divided into categories similar to a library system. Filo and Yang founded Yahoo! in March 1995 and solicited venture capital to fund the growth of their company. Thirteen months later, having risen to forty-nine employees, their initial offering of stock earned them a fortune. Yahoo! continued to grow, relying on a mix of links categorized both by hand and automated web crawling.

The web search engine business became extremely competitive in the late 1990s, and many of the larger search engines latched on to the idea of web portals. Web portals wanted to be the jumping-off point for users, a place that they always returned to (often setting up the portal as the default home page of their web browser) in search of information. Web portals offered a search engine, free web-based email, news of all types, and chat-based communities. By attracting users to their web portals, the web portal companies were able to sell more web-based advertising at higher rates.

Larry Page (1973–) and the Moscow-born Sergey Brin (1973–), another pair of Stanford graduate students, collaborated on a research project called Backrub, which ranked web pages by how many other web pages on the same topic pointed to them, using the ability of the WWW to self-organize. They also developed technologies to use a network of inexpensive PCs running a variant of UNIX to host their search engine, an example of massively distributed computing. In September 1998, Page and Brin founded Google Inc. The word *google* is based on the word *googol*, which is the number 1 followed by 100 zeros. Google concentrated on being the best search engine in the world and did not initially distract itself with the other services that web portals offered. In this, Google succeeded, quickly becoming the search engine of choice among web-savvy users because its results were so accurate. By the end of 2000, Google was receiving more than 100 million search requests a day.

In 2001, Google purchased the company Deja News, which owned a copy of the content posted to USENET since 1995, some 650 million messages in total. This became one of the many new Google services, Google Groups. The success of Google became apparent as a new meaning to the word rapidly emerged, its use as a transitive verb, as in "She *googled* Sergey Brin and realized he was born in 1973." From the beginning, Google was a big supporter of open-source programs, especially since their search engine servers ran a Google-modified version of the Linux open-source operating system.

## THE OPEN-SOURCE MOVEMENT

Open-source software is software that was published with the actual source code included instead of being in a format that only allows execution of the program. For example, when you purchase Microsoft Office, you can install and run the program, but you are not given the proprietary source code that allows you to alter the program and change it according to your own needs. Most open-source programs are distributed for free, with programmers contributing their time because they feel passionately about the programs.

In the 1970s, software became so important that new companies sprang up solely to produce software. Software became big business, spawning the economic giant of Microsoft and a host of other companies, and proved to be wildly more profitable than hardware, with profit margins of up to 80 percent. Because software was prone to be buggy, legal contracts called *end user license agreements* (EULAs) became common to prevent customers from suing because the software did not perform as advertised. Because the software industry did not actually sell their products, but just licensed their use, and because of the EULAs, the software industry was unique in being able to sell defective products with minimal legal consequences.

Operating system software during the 1970s also became more sophisticated. While most operating systems remained proprietary software, running only on the hardware also sold by the manufacturer, AT&T Bell Labs developed an operating system called UNIX in the early 1970s. UNIX became popular on university campuses and was unique in that the same operating system soon ran on many different hardware platforms, regardless of the manufacturer. UNIX also became the operating system of choice for the engineering and graphics workstation market that flourished in the 1980s and 1990s.

Richard Stallman (1953–), a programmer at the MIT Artificial Intelligence Lab, enjoyed sharing software that he wrote with other users and using the software that they shared with him. In 1984, inspired by his ideal that software should be free, Stallman refused to join the burgeoning software industry but quit MIT to create the GNU project. GNU is a recursive pun, meaning GNU Not UNIX. Stallman's crusading project aimed to recreate the UNIX operating system and the common tools on UNIX from scratch so that the GNU programs would be free from copyright and anyone could use them. Stallman and some law professors created the GNU General Public License (GPL), which Stallman characterized as a "copyleft" scheme rather than a copyright scheme. All GPL'd software must be released as free software with source code included. The software is not in the public domain but remains copyrighted so that it may not be readily used in commercial software. The GPL is characterized as "viral" in that any new software that includes any GPL'd software within it must then be released as GPL software. The GPL is not the only open-source license, but it is widely used.

In 1985, Stallman and like-minded programmers created the Free Software Foundation (FSF), funded through sales of GNU programs and donations. By selling software, the FSF seemed to violate its own philosophy of developing free software, but it was really just selling the disks and manuals, not the software, which, since it is licensed under the GPL, could be legally passed on to anyone else. The FSF created many of the common tools on UNIX, such as a compiler and text editors, but it remained for Linus Torvalds (1969–), a graduate student in Finland, to create the Linux operating system in 1991, a clone of UNIX that used the GNU tools. Other open-source UNIXes also appeared, such as FreeBSD and NetBSD. Open-source projects often concentrated on the UNIX operating system, but many projects were quickly ported to other operating systems, including the Microsoft Windows family.

While the founding of the FSF was a reaction to the commercialization of software, few people shared Stallman's idealistic vision, and copyright and patent rights were not abandoned by the open-source movement. In the 1990s, Linux vendors such as Red Hat and MandrakeSoft figured out how to make money from open-source software by selling convenience, service contracts, and consulting. Other vendors, such as MySQL, which marketed an open-source database, provided a choice to customers. If a customer wanted to use the MySQL database software as part of another software product, then the customer must either pay a licensing fee or also make their product free and open source.

Open source grew quickly during the 1990s and early 2000s and became an important market force. Open-source software was created in three ways: amateur projects, important projects sustained by competent professionals donating their time, and projects supported by companies who found that open source gave them certain competitive advantages. Many times, if the product was important enough, volunteers and company employees worked on the same open-source project. Examples of this include Linux, the GNU tools, and the Apache web server.

The open-source movement represented more than idealism and programmers doing what they loved. Like other inventors, once programmers saw that an idea was possible, they were halfway to replicating that idea. That meant that the makers of proprietary software were forced by competition from similar open-source products to continually improve, selling products with better features because the older features were becoming part of the common reservoir of software ideas.

Open-source projects became possible because of the internet. Spread around the world, open-source programmers communicated via networks, bound only by their common interests in a project and a quest for excellence. Projects were usually driven by a small group of core programmers, and decisions were usually made by consensus. Programmers rose to leadership positions because of their recognized skills, a mark of high-technology personal status and a form of meritocracy. Some programmers, called "über-programmers" were often many times more productive than the average programmer. They were the cream of the elite and the backbone of both important proprietary and open-source projects.

As programmers became more experienced with solving a particular type of problem, they created ever more sophisticated algorithms, which are like recipes for getting programs to do things. The community of software developers came to understand certain types of software so well that writing that type of software became a straightforward process. Text editors, operating systems, word processors, office productivity suites, email clients, media viewers, web browsers, and web servers all had solid open-source products by the early 2000s. This was called the *commoditization of software*.

Ironically enough, considering that open-source programs were often not backed by the financial resources of a large corporation, open-source software achieved a justly deserved reputation for being less buggy and more reliable than proprietary commercial software. This strange paradox occurred because software engineering researchers had found that one of the best methods to remove bugs from a software product was code review, where multiple programmers looked over the code and visually inspected it for flaws. The open-source process, by its very nature, where anyone can look at the code, had more eyes on the code than comparable proprietary software teams.

Programmers involved in open-source projects were driven by a passion for excellence. Open-source programmers put technical merit before market success. Unlike commercial companies, who shipped products to meet a deadline, open-source projects often only declared a product ready when it was really ready. Early releases of open-source products happened, but they were publicly labeled as *alpha* versions, not complete products. High-quality software was intrinsic to the open-source movement, though the movement remained littered with projects that had been abandoned because programmers lost interest. This was not a serious problem, as these products had become obsolete or did not attract sufficient users to maintain programmer excitement, a form of natural selection. When a product type was ripe to be commoditized, an open-source version appeared.

Open-source products presented a seeming paradox for computer security as well. Some pundits argued that because the source code was visible to everyone, malicious hackers could work their way through the code line by line, finding a bug or an oversight that could lead to a way to compromise the open-source product. This fear was certainly true, though experience showed that flaws in closed-source products were, again, more common because not as many programmer's looked at the source code for the product. After a period of seasoning, open-source products were more secure.

The open-source Apache web server was used by the majority of the world's web servers by the mid-2000s. Figures from the computer industry research firm Gartner in the early 2000s showed that Linux was the fastest-growing operating system in the world. IBM and Hewlett-Packard embraced Linux and other open-source projects and were already making billions of dollars off this change by the early 2000s. Apple Computer's operating system for its Macintosh computers, Mac OS X, based on the NeXT technology, debuted in 2001 and used an open-source flavor of UNIX as its foundation. Apple's later iOS, used for the iPhones and iPads, also had an open-source heart, though proprietary software lay on top.

Linux and the other open-source operating systems mostly ran on Intel-based microprocessors because they were comparatively cheap to purchase and support. By 2003, more Linux servers were shipped than servers with proprietary UNIX operating systems, part of a process of open-source operating systems coming to dominate the UNIX market and forcing the proprietary UNIX products out of the market, such as Solaris from Sun Microsystems, AIX from IBM, and HP-UX from Hewlett-Packard. Linux and other open-source products were making their greatest strides in the enterprise arena in the server rooms of large and small businesses. Though open-source desktop products were growing stronger, they had not reached prominence in the minds of common computer users in the early 2000s.

As noted earlier, Bill Gates stepped aside as CEO of Microsoft in 2000, though he remained as chairman of the board and was heavily involved in the company. He stepped down from chairmanship in 2014. Philanthropy became ever more important to him through the Bill & Melinda Gates Foundation that he and his wife had founded, and he gradually pulled back from the company he had founded, even leaving its board in 2020. The new CEO, Steve Ballmer (1956–), had joined the company early in its history and ran its marketing efforts. He ran the company from 2000 to 2014.

Even though Microsoft Windows and Microsoft Office ran on about 90 percent of the world's desktop computers in the early 2000s, Ballmer identified Linux and open-source projects as a major risk to Microsoft's future business. Microsoft was founded on the premise that people will pay for PC software, and early PC hobbyists in the mid-1970s resented the small company for charging for their products when everyone else gave their products away as part of the old hacker culture. Microsoft proved to have found the correct paradigm for the next three decades, but open source promised an alternative paradigm. Open-source software became increasingly important, driven by the enthusiasm of programmers. Linux and other open-source projects continued to grow because there was no company to fail and no profit margin to maintain. Google data centers ran on a tailored version of Linux, and its operating system products, Android and Chrome OS, were both built on top of Linux.

Satya Nadella (1967–), who was born in India, became the new CEO of Microsoft in 2014, succeeding Steve Ballmer. Nadella came from Microsoft's cloud and enterprise groups and had a different attitude toward open-source projects. He embraced open source, and Microsoft employees soon began contributing to open-source projects as part of their job duties. While selling software was still important for Microsoft, the computer industry had moved toward selling computing services and building content for the internet. In 2016, Microsoft included a Linux subsystem inside its flagship operating system, Windows 10. Over the years, this subsystem became a full version of Linux running in a virtual machine. The older attitudes of hostility at Microsoft toward open-source products had ended, as Microsoft found that coexistence led to more revenue from enterprise customers. In 2018, Microsoft acquired GitHub for $7.5 billion, the largest repository of source code in the world and a major repository for open-source code. SourceForge, founded in 1999, also remained a major repository for open-source code. The ideological struggle between proprietary software and open-source software had abated.

# SEVEN

# Moore's Law Triumphant

## PERVASIVE COMPUTERS

As the twenty-first century dawned, computers and electronics had started to become pervasive in everyday life. What was once high technology had become mundane. The increase of transistors per microchip and decrease in cost had continually led to a greater likelihood that electronics would find themselves embedded in more and more technology. Combined with the increased ability of data transmission, over wires and wirelessly, it also meant the increased likelihood of providing computer services online and more computers in people's lives.

## CONSUMER ELECTRONICS—MUSIC

The first electronic devices found in most homes in the United States were radios. Starting in the 1920s, mass-produced radios used vacuum tube technology to bring the growing number of amplitude modulation (AM) radio stations to Americans. By 1921, stations also began to broadcast shortwave signals to attract an even larger worldwide audience.

After the invention of the transistor in 1947, receiving radio broadcasts went mobile. Billions of transistor radios were sold after the first one, the Regency TR-1, was jointly created by Texas Instruments (TI) and Industrial Development Engineering Associates in 1954. Raytheon, Zenith, RCA, and Crosley soon followed with their own models. Chrysler put the first radio in a car in fall of 1955. Masaru Ibuka (1908–1997), founder of the Tokyo Telecommunications Engineering Corporation (later renamed Sony), took advantage of Japan's lower labor costs to drive miniaturization, improve quality of components (transistors were notoriously flakey and had only a 20 percent effectiveness rate when first introduced), and lower prices. Integrated circuits allowed for even greater miniaturization. Radios became so inexpensive that by the 1970s, they were embedded in everything from pretend Coke cans to stuffed animals and sometimes given away as promotional items. Transistor radio sales were driven by the period of prosperity following World War II that had created a large generation with leisure time and disposable income.

Music and portability in the late 1970s saw the rise of boom boxes, like those created by Morantz and General Electric (GE), and portable cassette players, like the Sony Walkman (released in 1979), which allowed consumers to not only listen to the radio but also their own collection of music. These devices benefited from the cassette tape recorders of the 1970s that allowed consumers to record their own or a friend's 33 1/3 revolutions per minute (rpm) long-playing vinyl phonograph records (called LPs or

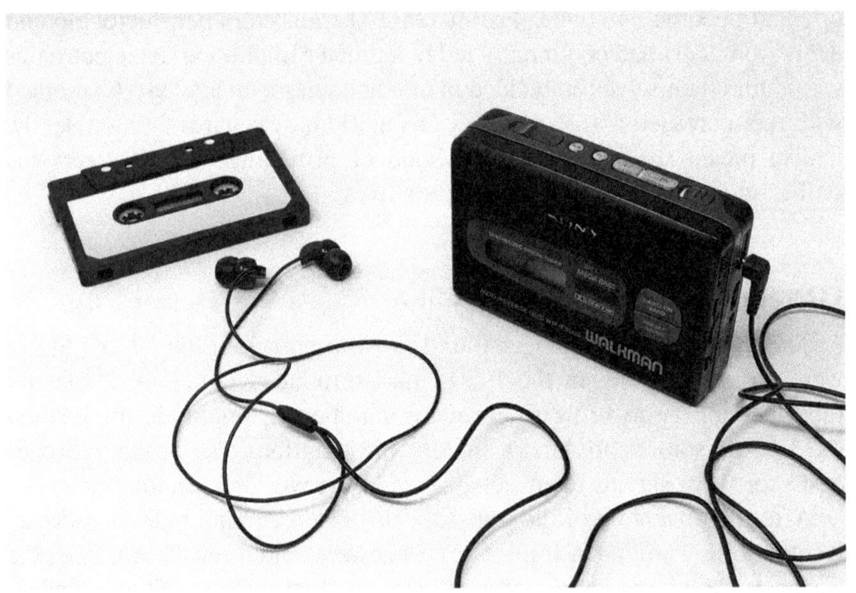


Cassette, headphones, and Sony Walkman. The transistor radio had created a portable means of listening to entertainment, and the Walkman went a step further in allowing the user to control content at the user's discretion. (Massimo Scacco/Dreamstime.com)

albums) to cassette tapes. The Sony Walkman was oriented toward individual listenership with earphones and became part of the jogging craze, while the less private boom box became associated with hip-hop and breakdancing in the 1980s.

A major move from analog to digital arrived with the Sony Discman in 1984, which utilized the compact disc (CD) optical storage technique developed in 1982. The discs could hold eighty minutes of uncompressed audio or 700 megabytes (MB) of digital information, which, in 1982, was far more than the typical computer hard drive at 10 MB. This drove the use of optical discs for storing data and distributing software for personal computers (PCs). The Discman also helped drive the transition from vinyl albums to CDs for the music-buying public, and vinyl began disappearing from retail stores in 1988. CDs were also cheaper to produce and distribute for the record companies. Ironically, in 2020, vinyl albums outsold CDs. CDs had lost out to online music sources, but audiophiles and young people rediscovered the rich sounds of the older technology and the joy of holding the music in your hands.

British inventor Kane Kramer (1956–) has been credited with the invention of the digital audio player, the first being called the IXI. Four prototypes were created from the original in 1979. It used an LCD screen, control buttons, and a solid-state bubble memory of 8 MB that could hold approximately 3.5 minutes of audio. The patent issued in 1987 lapsed when he failed to renew, and the design entered the public domain. This benefited Apple between 2006 and 2008. It successfully argued prior art by Kramer when seeking to declare Burst.com's patents invalid while defending its Apple iPod. Burst had earlier won $60 million in a settlement with Microsoft in 2005.

The portable digital audio player took advantage of the programming chops of computer scientists around the world who created audio coding formats—especially ones that compressed music to take up less memory without losing audio fidelity. One of the most popular, released in 1993, the MP3, is a "lossy" algorithm. It has acceptable but reduced fidelity. Because of its ability to reduce files to a tenth of their original uncompressed size, however, it became the standard for many players. The MP3 player became synonymous with a digital audio player in the public's mind.

The first and most successful player in the U.S. market was the Audible Player from Audible.com in 1998. It had no display and half the memory of Kramer's 1979 prototype. However, using a rudimentary proprietary audio compression, it could hold two hours of music. In the fall of 1998, the Diamond Rio (PMP300), with 32 MB, also arrived and sold well that holiday season. It also used the MP3 format. Early audio players had small hard drives to store music (or insertable disk media, like the Iomega Clik! drive of 1999), but most evolved to use solid-state storage.

Numerous MP3 players entered the market. Users realized the value in converting their own music to MP3 format on their computers and loading that onto their players, but more users, with MP3 as a standard format, began sharing songs. File sharing across the internet had existed since its origins and had been augmented by systems like USENET. In 1999, the website Napster, created by Shawn Fanning (1980–) and Sean Parker (1979–) made it incredibly popular and easy to share music. The site existed for two years and had more than twenty-five million users. Some musicians argued that Napster increased exposure to their music, but other musicians and the recording industry forced Napster, through a successful U.S. district court decision, to stop the sharing in 2001. Napster filed for bankruptcy in 2002. All the copycat music sharing sites that sprang up also eventually shut down. Eventually, even sites that only allowed freely listening to music but not sharing files, like Grooveshark, shut down for the same reason.

In January 2001, Apple introduced the iTunes program to manage music files on a desktop computer, which initially only ran on Macintosh computers. The origin of iTunes was SoundJam MP, created in 1998 and purchased by Apple in 2000. Apple released the iPod music player more than eight months after iTunes. It had 5 gigabytes (GB) of storage on a hard drive and was roughly the size of a pack of cigarettes. With a clever touch interface and ease of use, iPods became enormously popular, dominating the market for music players. Numerous iPod models, some much smaller, barely bigger than a pack of matches, were released. Fifty million iPods were sold in 2008, the height of the market, and over 400 million units in total were sold. The development of the iPhone and other smartphones considerably shrank the market for separate music and media players. In 2022, Apple was still manufacturing the iPod Touch—first released in 2007—essentially an iPhone without a cell connection.

In 2003, iTunes introduced the iTunes Music Store, having managed to get major music labels to agree to sell individual songs for 99 cents. To convince the music labels to cooperate, Apple had to provide a digital rights management (DRM) process to limit the ability of users to share purchased songs. In the first six years, iTunes sold over six billion songs, and in 2009, Apple dropped DRM. The music business had changed so dramatically that DRM was no longer needed. The iTunes program became a clearinghouse for purchased music and music downloaded through other means. Apple pushed iTunes to manage photos, iPhone activation, podcasts, apps, videos, and e-books, turning the program into unwieldy mess by 2010. In 2019, Apple eventually decided to break up iTunes into separate programs for separate functions.

## CONSUMER ELECTRONICS—VISUAL MEDIA

Because of computers, how Americans watched visual media changed even more than how they listened to music. Prior to the advent of video recording machines in the 1970s, consumers watched movies during their first run in the theaters. If they missed that, they had to rely on the whims of second-run theater houses or watch films several years later on television, with the movies heavily edited for content and length to accommodate numerous commercial breaks. Some really popular movies, especially Disney movies, regularly returned to theaters. You watched television shows when the television networks ran them. The three big television networks, CBS, NBC, and ABC, ran reruns of that season's shows in the spring and summer when the regular season had run its course. After that, you were typically out of luck until, local UHF (ultrahigh frequency) channels started running old shows. A few shows, like *Star Trek*, actually benefited from this second life, gaining far more fans in reruns than when originally shown. For the most part, unlike music, people did not own studio-produced visual media other than a small cadre of cinephiles who owned mostly 8 mm versions of the original 35 mm and 70 mm reels, and the projectors to show them.

The videocassette recorder (VCR) industry began with AVCO, a company founded in the 1920s that focused on aircraft equipment. AVCO purchased the Embassy Pictures studio in 1967 and targeted bringing movies to the home with the Cartrivision machine. AVCO created a system that allowed for recording television shows when they broadcast to watch later (a process called *time shifting*), renting Hollywood features on videotapes, and owning some instructional tapes. Unfortunately, only one other major Hollywood studio, Columbia Pictures, joined AVCO in making videotapes of movies available. Hollywood was fairly skeptical of changing its model of movie distribution. The limited introduction of Cartrivision stumbled because the devices were large, expensive, buggy, and too complicated for retail salespeople to understand and sell. AVCO was on the verge of a national launch with Sears when its executives saw the Sony Betamax on a business trip to Japan. The company quit the market.

Sony focused on the time shifting approach with its Betamax. Original studies suggested that customers might own up to five tapes for recording television shows. The market was much larger than the company realized. Customers wanted to store many recorded events, including working with video cameras (a process begun by hobbyists) to store their own creations. VCRs improved gradually as users found ever more reasons to buy more blank videotapes.

Sony started selling its Betamax machines through stereo, camera, and electronic equipment stores, and they sold well in the late 1970s. None of the Hollywood studios agreed to distribute movies with Sony, though the equipment resellers and others began selling prerecorded content. Illegally recorded movies, old movies in the public domain, and pornography also had roles in the device's acceptance. In 1976, when businessman Andre Blay (1937–2018) cut a deal with 20th Century Fox to distribute movies at least four years old on videotapes, Americans, for the first time, could easily own movies (although still expensively), and Blay used the possibilities of owning movies (including through the Video Club of America mail-order house) to sell the Betamax machines themselves.

While AVCO and Betamax began the VCR revolution, another Japanese company, Victor Company of Japan (JVC), partnering with RCA in the United States, won what was called the "videotape format war" with its Video Home System (VHS). The Betamax had higher technical quality but suffered from shorter record and playback times. The shorter times could have played a role in the Betamax's demise, but the refusal of Sony to license its technology to other manufacturers kept its VCR machines more expensive. Meanwhile, VHS technology was widely licensed, and competition drove down the cost of the machines. Other Hollywood studios began to offer their movies on expensive prerecorded videos for the rental market. RCA also drove an effort to dominate the prerecorded tape market, which then drove the purchase of VHS machines by consumers, which drove the format of prerecorded movies for rent and purchase, continuing a feedback loop. What business people call *network externalities* proves the importance of standards for the marketplace. Prerecorded material was the killer app for VCRs.

In the late 1970s and early 1980s, small video stores purchased prerecorded movies to rent out and usually VCRs to take home overnight as well. These videotapes usually cost the rental stores $80 to $100, priced in the expectation that the store would recover its costs and make a profit before the tape expired from use. With prices like that, most consumers did not purchase their own tapes, but many consumers started to buy their own VCRs as those prices came down. When the hit movie *Top Gun* (1986) was offered at a much lower price point and sold an astounding $40 million in sales, Hollywood realized that they had a new market for their products. People started to accumulate VHS movie libraries, which they dutifully replaced with DVDs once that technology came out. Small video rental stores succumbed as large chains took over, with the introduction of Blockbuster in 1985 eventually peaking at over 9,000 stores in 2004. Blockbuster later succumbed to the rise of Netflix DVD mail rentals and Redbox kiosks in stores that rented a small selection of the most popular videos. A single proud Blockbuster store remained in 2022.

VCRs mostly disappeared when two separate devices took their place. Digital Video (or versatile) disc (DVD) optical disc players, introduced in 1996, had higher-quality playback of prerecorded material and contained 4.7 GB of digital data on single-layered discs and twice that capacity on double-layered discs. DVD recorders later became common for data storage on computer systems, supplanting recordable CD technologies. Digital video recorders (DVRs), like TiVo, introduced in 1999, were computers with a large hard drive attached to your television; they recorded programs when they aired on broadcasts over the air, cable, or satellite networks and allowed you to watch them time shifted at your leisure. People had been doing the same with their VCRs, but videotapes held a limited amount of content. DVRs held much more content at a much higher quality. While a VCR was a basic computer, DVRs were full-fledged PCs devoted to a single task.

Marc Randolph (1958–) and Reed Hastings (1960–) founded Netflix in 1997. They began by mailing DVDs but pivoted to include streaming, or video-on-demand (VOD), in 2007, as internet speeds had increased to the point where it had become viable. Netflix executives realized that content providers—movie studios and television production companies—would eventually pull their content from Netflix and create their own streaming services. As a technology company, Netflix could raise lots of capital, and the company poured unprecedented amounts of money into creating its own series and movies. Amazon and Apple followed suit in producing content, and other streaming services became common in the late 2010s. The phrase "direct to video" was no longer pejorative with major stars and multimillion-dollar budgets joining the transition. Netflix had 183 million customers worldwide in 2020. Computers and the internet had completely changed the delivery model for entertainment. Streaming content directly at the customer's discretion had come to dominate the media landscape. Many customers "cut the cord," dropping cable and satellite services in favor of streaming. The COVID-19 crisis beginning in 2020, which saw many people stuck in their homes, only accelerated that transition.

## TOYS AND GAMES

The first electronic toys were radio-controlled (RC) vehicles principally oriented toward adult hobbyists, especially given the cost. Nikola Tesla invented an RC boat in 1898. He amused himself by using the boat to trick people into thinking they could control the vehicle by shouting commands. Mid-1966 saw the first RC car from the Italian company Elettronica Giocattoli, a 1:12 scale Ferrari. With the shrinking size of electronics in the 1970s, electronics were found embedded in toys of all types. The National Semiconductor Quiz Kid game, introduced in 1975, tested children on basic arithmetic operations. The electronic game Simon, released in 1978, tested children's memories of a sequence of four lighted panels.

In 1999, Sony introduced a consumer model robot dog named AIBO that used a 64-bit processor, camera, microphone, speaker, and touch sensors on its feet and head; it also had a moving head, tail, and legs. Costing thousands of dollars discouraged sales, but it was recognized for its advanced approach to consolidating electronic technology and was placed in the Carnegie Mellon Robot Hall of Fame in 2006. LEGO introduced the MINDSTORMS platform in 1999 with a programmable brick and controllable sensors and motors. Working with inventor Dean Kamen (1951–), the company helped create the FIRST LEGO League worldwide competition. In 2019, the competition had over 300,000 participants aged nine to fourteen years old.

The shrinking size and price of electronics allowed for new types of game consoles. The first true totally digital handheld electronic game was Mattel's *Auto Race*, released in 1976. It displayed a limited number of characters on a small screen. Other toy companies, like Milton Bradley, also released single-game handheld devices. Nintendo brought out the Game Boy in 1989, a handheld 8-bit console that could play numerous games by inserting game cartridges. It quickly sold a million units within weeks of release by taking advantage of selling versions of already popular games,

LEGO robot ready for a FIRST LEGO League competition. The LEGO controlling computer "brick"—in the center—introduced robotics and computer programming to millions of children. (Peanutroaster/Dreamstime.com)

like *Super Mario* and *Tetris.* It also had better battery life than its competitors.

Game consoles (both handheld and in the home) continued to evolve and became, especially in the home, much more than simple game platforms by taking the place of DVD and Blu-ray players and cable set-top boxes. By using the internet, Microsoft's Xbox Live, released in 2002, offered owners of the Xbox access to single-user and multiplayer online games and other media. Home consoles added more than buttons and knobs as a user interface in the mid-2000s. Microsoft released the Kinect system in 2010. It evolved to utilize cameras, infrared sensors, and microphones to allow gestures and speech recognition to interact with games like *Space Invaders Extreme* (a variation of the 1980s *Space Invaders* game), *Kinect Star Wars,* and *Dance Central.* The system became so sophisticated and comparatively cheaper than other equipment that academic researchers began using the system. In 2020, Microsoft counted nongaming applications as the biggest market for the system.

In 2006, Nintendo offered the Wii system. The Wii included two remotes, one for each hand. One of the remotes connected to the console with wireless Bluetooth and used accelerometers and infrared to determine its location in three-dimensional space. The control also had a vibrating "speaker" that gave haptic feedback, a "rumble," to the user. Other interfaces were also sold, such as a vinyl dance pad for playing 2009's *Just Dance*, a drum and guitar set for playing 2005's *Guitar Hero*, and, for nongamers, a balance board for 2009's *Yoga Wii*. Numerous people injured themselves or others playing many of the Wii games because of the physical requirements. This included tennis elbow from playing *Wii Sports*—a condition that became known as "Wii-itis." The Wii sold more than 100 million units worldwide. Electronic gamification became part of home exercise in the 2000s in no small part because of the Wii. For example, the Peloton stationary bicycle, launched in 2012, came with a twenty-two-inch touchscreen and included canned and live online classes.

## THE ROAD TO SMARTPHONES—THE PDA

What people began calling a *smartphone* in the 2000s is an amazing example of technological convergence. Electronic organizers, calculators, dictionaries, maps, games, clocks, calendars, personal digital assistants (PDAs), pagers, email devices, cell phones, GPS, miniaturized electronics, improved energy use and batteries, and ubiquitous network technologies all combined to create the electronic Swiss Army knife of the twenty-first century. The term *smartphone* had been used in the 1960s, sometimes to describe landline phones with special features, but one device, the iPhone, completely changed how people thought about phones.

In the 1980s, pocket-sized computer/calculators began adding features and programming languages so that users could create their own tools. The programming of calculators first began with programming keystrokes to do operations. The language of choice was often BASIC. Pocket computers, like the $249 TRS-80 Pocket Computer and Sharp PC-1211, both introduced in 1980, could store 1,424 bytes and had a one-line twenty-four-character screen and a full QWERTY keyboard. Some owners figured out how to use them to store up to 100 personal contacts.

Electronic organizers, called personal digital assistants (PDAs) or personal information managers (PIMs), extended personalized calculation devices to nonnumerical data. In 1975, Satyan Pitroda (1942–) filed the first patent for an electronic diary; it had the shape and look of a calculator but with letters as well as numbers and a single-line screen. Many manufacturers, including Psion, Casio, Sharp, Hewlett-Packard, and Tandy/Radio Shack, entered this field in the 1980s, aiming to put the physical Rolodex—a staple on the desks of the business world for storing addresses, phone numbers, and other contact information since the late 1950s—into businesspeople's pockets.

Electronic organizers started humbly. The Psion Organiser, released in 1984, had a calculator, simple data storage, clock, and a single-line monochrome screen. The Casio SF-2000 had two lines on the screen and boasted it could store over 533 items, including addresses and notes, in its limited 2K memory. By the late 1990s, the Casio Business Navigator BN-40A had 4 MB of available writable space and a cursor-controlled or touchscreen graphical interface; it was able to synchronize with a PC to backup data using Casio software. This clamshell-style of PDA was manufactured into the 2000s because it was comparatively inexpensive as a nonnetworked device.

## THE ROAD TO SMARTPHONES—MOBILE COMMUNICATION

Two-way electrical communication in the mid-twentieth century required either wires or point-to-point "walkie-talkie" transmitter/receivers that used radio signals. Fairly unusual instances of a radio being hooked to the phone system occurred, sometimes in as rudimentary a fashion as a radio and a phone receiver being held together. The first commercial public use of wireless communication connecting to the ever-ubiquitous phone system so that an individual could call any connected phone in the United States took the form of a car phone. First introduced in 1946 by Motorola, by linking to Illinois Bell Telephone, the customer's equipment weighed eighty pounds—needless to say, too heavy and too power hungry to be carried by hand. It connected to distributed radio antennas using the mobile telephone service (MTS) standard. It had limited appeal given the cost ($15 a month) with customers only in the thousands in the 1940s.

Enough demand existed to continue the approach. Alternative services gradually cropped up worldwide using different standards. Great Britain saw the Post Office Radio Phone Service in 1959. AT&T introduced a system called improved mobile telephone service (IMTS) in 1965, an automatic system that did not require an operator. Continental Europe, the

Soviet Union, and the Eastern bloc also had such systems. A consumer could purchase a car phone until 2008. Even in 2008, however, while progressively smaller and lighter, they used a completely different telecommunication system than the cell phone system that replaced it.

The first cell phone call occurred on a street in New York City in 1973 in front of a crowd of journalists during a telecommunications business conference. Martin Cooper (1928–), a member of a team put together by Motorola's chief engineer for mobile communications, John F. Mitchell (1928–2009), made the first public phone call on a handheld device: the DynaTAC 8000X (short for Dynamic Adaptive Total Area Coverage). Cooper dialed the number of the head of a rival team at AT&T Bell Labs, Joel Engel (1936–). Cooper claimed cartoon character Dick Tracy and his wristwatch radio as his inspiration to create mobile telephony. The phone utilized an analog cellular antenna that had been set up on the roof of a nearby building for the demonstration, which connected him into the AT&T system of land phone lines. Cooper gently mocked his competitor that Motorola was first to demonstrate a handheld personal mobile telephone.

It took Motorola more than ten years to bring the product to market in 1984. The phone weighed 2.5 pounds and was labeled "the brick" because of its approximate size and shape. It had enough battery life for half an hour of use. With a price of $3,500, there was a limited market. And making the phones was just one of the hurdles. A network of cell towers had to be built to carry the radio signals, regulatory permission had to be obtained for radio spectrum, and contracts had to be negotiated to hook the cell towers into the current telephone system. A cellular network was created to serve hexagonally shaped geographic areas, each hexagon served by a directional antenna. This approach had been invented by engineers Douglas Ring (1907–2000) and W. Rae Yong (1915–2008) at Bell Labs in 1947, almost forty years earlier. After the 1984 introduction, it took Motorola yet another ten years to solve all the problems of handing off calls between cells, multiple use of frequencies to utilize the limited frequencies available, and creating the relatively inexpensive transmitters that could utilize existing structures that drove cellular's adoption. By 1990, there were a million cell phone users in the United States.

In 1989, Motorola had released its analog successor, the MicroTAC. At only twelve ounces and with a pop-up antenna and a "flip" receiver that revealed its keypad, it presaged the cellular phones of the 1990s. By 1994, this phone's successor, the MicroTAC Elite, weighed only four ounces and had become GSM-compatible to allow for use in European cell networks.

It offered a built-in message machine (voice mail) and phone directory—both firsts. The Elite also utilized, for the first time in a phone, lithium-ion batteries. Motorola ruled the 1990s cell phone world, and even had a real hit with its Razr phone in 2004, but lost its claim to largest in the market in 1998 to a company in Finland known to many Finns for its rubber boots: Nokia.

Nokia began in 1865 with Fredrik Idestam (1838–1916) building a riverside pulp mill in the small town of Tampere, Finland. The company moved operations to Nokia, Finland, in 1871. In the next century, the company expanded to numerous forestry products, power generation, rubber products, cables, and electronic equipment, including radios, computers, and televisions. In 1982, Nokia had marketed the Mobira Senator car phone using the NMT system and the first handheld phone, the Mobira Cityman 900, in 1987. Nokia and Ericsson, a Swedish networking and telecommunications company founded in 1876 as a telephone company, drove the Nordic countries to introduce the Nordic Mobile Telephony (NMT) system in 1981 to replace the competing systems of Finland, Norway, and Sweden. As a collaborative network, NMT became the first mobile phone system to allow international roaming.

In the 1990s, competing systems existed in the United States and Europe, the two biggest markets at the time. The United States had multiple standards, including the code-division multiple access (CDMA) standard (as a point of interest, the approach had been utilized experimentally in 1957 in the Soviet Union and later used for their Altai vehicle mobile phone system). Europe had developed the global system for mobile communications (GSM). Finland launched the first GSM network in 1991, and that precipitated its place in cell phone history. Nokia released the first GSM phone, the Nokia 1011, which ran on the Nokia-built network, and had it demonstrated by Finland's prime minister, Harri Holkeri.

Despite the early U.S. lead with Motorola's "brick" cell phone, further development suffered from larger political issues. The AT&T monopoly had created the world's best landline phone system, and the vast majority of Americans had access to a home phone. In comparison, many other countries lacked such a technological investment and were free to explore creating a new system that was far cheaper to build than miles of wires on poles. The Federal Communications Commission (FCC) in the United States, along with Congress, deliberated over whether AT&T should create the system or allow for competition. In 1974, the federal government filed the antitrust lawsuit that eventually broke up AT&T after ten years of litigation and negotiation. AT&T actually had a working cell system in Chicago in 1978, but technological inertia, the monopoly breakup, and multiple standards of competitors all contributed to slower installation and adoption. It was 1995 before the U.S. launched a second-generation cell communications system, while Europe had done that with GSM in 1991.

Innovations flowed forward from around the world. Japan's Nippon Telegraph and Telephone Corporation (NTT) introduced the first automatic analog cellular system in Tokyo back in 1979. The Short Message Service (SMS) became hugely popular because of the cost structure of sending texts versus voice. In Finland, this evolved into the first phone-based banking service in 1997. The first mobile phone payments for Coca-Cola vending machines was also introduced in 1997. Japan's Kyocera VP-210 Visual Phone had the first built-in camera in 1999, and Nokia first introduced the camera phone to American audiences with the Nokia 3600. Some cell phones contained simple games, such as Nokia's *Snake*.

Europe had settled on the GSM standard, and when that became the worldwide standard, Nokia gained ground by selling the GSM-capable phones and also the infrastructure for those phones. By the end of 1998, Nokia had sold 100 million cell phones worldwide, and one in three Finns owned a cell phone, the biggest percentage of a single nation's population in the world. Nokia accounted for almost a quarter of Finnish exports and employed over 130,000 people in 2000. It controlled 41 percent of the world cell phone market in 2007, and until 2012, it accounted for the largest percentage of phones manufactured. A rapid decline for Nokia set in as new technologies and market dynamics ended the Nokia hegemony.

## THE ROAD TO SMARTPHONES—"THE MELTING POT"

The first commercially successful tablet computer was the GRiDPad created by Jeff Hawkins (1957–) in 1989. The GRiDPad used a touchscreen and a stylus and was oriented toward replacing clipboard and paper forms. Anticipating the iPad of the 2000s, Hawkins even created the concept of flipping the screen from portrait to landscape, though 1993 saw the last release of the GRiDPad. Hawkins went in a different direction and used his experience to create the PalmPilot as a more intuitive organizer—the first commercially successful general-purpose touchscreen PDA you could hold in one hand.

Hawkins started Palm Computing in 1992. He hired Donna Dubinsky (1955–) and Ed Colligan (1961–) and wrote handwriting recognition software for other devices: Casio's Zoomer and the Apple Newton. In 1996, and now owned by U.S. Robotics, Palm created the PalmPilot—later simply called Palm after a legal tussle with the Pilot Pen Corporation. At approximately three-by-five inches and three-quarters inches thick, it came with programs for a calculator, address book, notepad, and more on a grayscale screen. Future generations added color and games. They also added a camera, MP3 player, and Wi-Fi through a SD slot in the Palm Zire 71 in 2003. Hawkins, Dubinsky, and Colligan, concerned about the direction of the company, left in 1998 to create Handspring. Their new company then merged back with Palm in 2003 and brought with it what became known as the Palm Treo 600—the company's first exploration into the cell phone space—and thus helped to create the smartphone category. The Treo was an open platform that encouraged third-party development. With a software development kit (SDK) available, over 11,000 applications were created. Other companies, however, became more dominant, and the last Palm, the LifeDrive, came out in 2007 and sold poorly.

During 1992, while Jeff Hawkins was starting Palm, IBM debuted a device, code-named "Sweetspot," that utilized mobile networks at the annual COMDEX computer and technology trade show in Las Vegas, Nevada. This was a prototype of the IBM Simon Personal Communicator released in 1994 and created by a team led by IBM's Frank Canova (1956–). The device has been called the first smartphone, although not by IBM; it could send and receive emails and faxes. It also had a touchscreen and stylus interface and included calendar, appointment, address, world clock, and notepad applications. It also allowed for PCMCIA (Personal Computer Memory Card International Association—a group of international companies which established the standard in 1990) cards to add additional memory and third-party applications, although with no established app marketplace or development environment, only one fairly expensive application, DispatchIt, was created. The device weighed more than a pound and may have been ahead of its time. Articles expressed admiration for the technological integration, but IBM's BellSouth partner sold fewer than 50,000 units. It also used the analog first-generation cell network technology, which was not really designed for data. Second-generation (2G) and digital communication came too late to benefit from Canova's work. With a change in management at its headquarters, IBM abandoned the smartphone, and Canova left the company in 1994. He joined Palm in 1997 and worked on Palm's PDAs.

A year before Canova joined Palm, two high school buddies from Windsor, Canada, started a company called Research in Motion (RIM) and released the world's first two-way pager, the 850, using the DataTAC network. It could both receive and send texts using the cell network. RIM followed this two years later with the 950, which allowed email and had an eight-line screen, a scroll wheel for navigating, and a little keyboard. It included a calendar and to-do list, but text communication was definitely RIM's killer app. This newer pocket device was named BlackBerry by a marketing firm because the little keyboard looked, to their minds, like a strawberry, but BlackBerry sounded better.

RIM created relationships with cell carriers around the world, and by model 6210 (named in *Time* magazine's 100 All-Time Gadgets) in 2003, it had built-in phone capabilities. By 2006, it had earned the name "Crackberry" from users—a portmanteau of *BlackBerry* and *crack* cocaine because it was so addictive to use with its ever-available text, email, and *Brick Breaker* game. The QWERTY keyboard allowed people used to that interface to quickly type messages with their thumbs—faster than what the cell phones with their 0–9 keyboards allowed. That physical keyboard became its defining feature versus the smartphones to come. In 2009–2010, at RIM's peak, RIM controlled 20 percent of the smartphone market globally and 50 percent in the United States. RIM benefited from its system being more secure, and government entities continued to use BlackBerrys into the 2010s.

Nokia created the 9000 Communicator (in reference to the "communicator" in *Star Trek*) in 1996. Slightly smaller and lighter than Canova and IBM's Simon, it included a rudimentary web browser and an software development kit (SDK) for third-party developers—two elements that contributed to the future success of any smartphone. It also utilized a layered hardware technique that solved the problem of electronic component interference. The version in the United States did not sell as well as it did in Europe. It cost around $1,500 in 1996 and was not small enough to fit in anyone's pockets. Nils Rydbeck, the research and development director at Ericsson's Mobile Phone division in Sweden, created a similar device with the GS88 that was the first to be called a "smartphone." With an eye toward competing with Nokia, Ericsson decided to rethink the size and weight and held off marketing the GS88 and released the R380 in 2000 instead. Less than half the price and weight of the Nokia, having saved some weight by using a touchscreen keyboard instead of a physical keyboard, it sold well.

As smartphones developed, companies struggled to create user interfaces to manage web pages never designed for such small screens. Ericsson, Motorola, Nokia, and Panasonic bought into a company called Symbian in

1998 to compete against the BlackBerry and Palm interfaces. Symbian had spun off from the Psion company in London, which had played a role in creating software and hardware for handheld devices since 1980. The four companies considered partnering with Microsoft, which had been building a mobile platform since 1995 called Windows CE, but they decided they could more easily manage the much smaller Symbian arrangement. When Nokia later began dominating the direction of Symbian, the other companies fell out of the relationship. This was somewhat ironic given that they had invested in Symbian to counter the dominance of a large company. Nokia had become the dominant company in cell phones. Windows CE did become part of other handheld PCs from Casio, Compaq, Hewlett-Packard, and others, but Windows-driven handhelds never became a dominant part of the landscape.

One of the issues with trying to bring the Web to handheld devices lay in the nature of web pages, which had been built in hypertext markup language (HTML) for larger computer screens. Japan realized the third generation in wireless technology first in 2001, and one of the reasons for this was the unique dominance of NTT DoCoMo in the country. NTT DoCoMo created the i-mode mobile web service in 1999, and it was the largest company in Japan in 2001. The i-mode system used a compact HTML that made the mobile websites easier to recode for the smaller screens. Europe and America used WAP (Web Access Protocol). NTT DoCoMo's size encouraged the cell phone manufacturers and owners of web pages in Japan to augment their online offerings for i-mode. A quarter of Japan's population used the system by 2002. The system never sold well outside of Japan, however, because NTT DoCoMo did not play such an outsized role elsewhere.

Rydbeck of Ericsson termed devices like the R380 as part of "the melting pot" strategy, meaning multiple functions came together in a single unit—for example, having the phone take the place of a media player. The first phone to have a built-in MP3 player was Korea's Samsung SPH-M2100 in 1999. More MP3-playing cell phones were sold by 2006 than the total of all the stand-alone MP3 players. A year later, a billion MP3-playing cell phones had been sold. Apple saw its market in stand-alone audio devices under threat, and for that reason more than any other, it reexamined the market it had explored and failed at back in 1993 with the Apple Newton.

On January 7, 2007, in one of the most famous tech demos, Steve Jobs mesmerized the audience as he introduced the iPhone to the world with ninety minutes of superlatives and live demonstrations. Only five years earlier, he had rebuffed the Handspring founders, Colligan and Dubinsky, who had approached him about partnering. At the time, Jobs was still haunted by the Newton flop (which we explore below) and had told them he did not believe handhelds would be at the center of future computing. Meanwhile, Jobs had partnered with Motorola by putting an iPod inside the Motorola ROKR E1 cell phone, making it the first phone that could download music from iTunes. The look and feel of the phone so insulted Jobs's streamlined aesthetic sensibilities—he had famously lived in a house with barely any furniture and could be uncompromising about design—that he investigated having Apple build its own phone.

After the demo, and despite its high price, the iPhone was an instant hit. Apple's stock shot up. The iPhone initially debuted only on AT&T's EDGE system, which was slower than the third-generation (3G) systems then available, but it was fast enough, included Wi-Fi and Bluetooth for the home or office, and people loved the user interface. A very small team had secretly worked on the interface, which used a multitouch interface from a company called Fingerworks that Apple had bought in 2005. A great deal of work went into ergonomics. The immediacy of the response as people swiped, tapped, and pinched perfectly sized and colored icons that seemed three dimensional created an emotional connection to the user. The iPhone seemed alive and responsive. It had three sensors: proximity (shut off the display when holding it next to your ear), light (adjusted screen brightness to save power), and accelerometer (recognized how the device was oriented and moved through space).

Apple also learned the positive lessons of NTT DoCoMo with the web interface. In fact, Jobs claimed it was the "first usable browser on a cell phone." That was hyperbole, but for everyone living outside of Japan's cellular infrastructure, it was a revolution. The iPhone ran the Safari browser, which dynamically made existing web pages accessible on the small screen. Unlike with NTT DoCoMo's approach, those websites did not need to be individually reprogrammed.

The release caught the industry flat-footed, but Apple had to play catch-up as well. In typical Apple fashion, it had created a walled garden with no access for third-party developers. Jobs's inner circle had to convince him to create a gate in the wall. Six months into 2007, developers could create web apps for the iPhone, but web apps still could not take advantage of the iPhone's hardware. Jobs's inner circle pressed again, and an SDK came out in October 2007, followed by the Apple App Store opening a year and a half after the iPhone's introduction. Jobs still insisted that any app developed for the store had to be approved by Apple for distribution so that Apple still retained some control over quality.

Inspired by the interface and sensors, developers rushed to create programs for the iPhone, and the App Store cemented the iPhone as the technology leader with games and tools. Apple sold 1.39 million iPhones in 2007 and ten times that many in 2008, and sales continued to accelerate until reaching a peak of 231 million iPhones sold in 2015. iPhones drove Apple to a trillion-dollar valuation on the stock market and constituted the majority of the company's revenue.

Another large company had been secretly looking into a smartphone and took a totally different, non-walled-garden, yet ultimately successful approach: Google. In 2005, Google purchased a company named Android Inc. for $50 million from Andy Rubin (1963–), who had started the company after leaving Danger Research Inc. That company had created the Sidekick—a fairly powerful smartphone marketed (or mismarketed, according to its inventors) through T-Mobile Communications. The Android operating system was based on open-source Linux, unlike Apple's, which was based on its proprietary OS X. It aimed to bring everything that people had familiarity with on their computers to the phone: Gmail, Google Search, Google Maps, Google Calendar, and so on. Google's first smartphone, the G1 (powerful but deemed ugly by many in the press), went on sale in October 2008 and sold much slower than the iPhone had, though it did take good advantage of the fourth-generation (4G) network and the Long-Term Evolution (LTE) standards.

Despite the initial slow sales of the G1, Apple and Google recentered the United States as the main player in smartphone design. The operating system became central to the smartphone ecosystem, and Silicon Valley led the world in software development. Ericsson left the mobile phone world by selling off its mobile unit to Sony in 2011. Nokia had pushed Symbian away from touchscreens and so needed to pivot to compete against Apple. It dropped Symbian and adopted Windows in 2011, but Windows suffered from an image problem as an also-ran with its interface. Microsoft actually bought Nokia's smartphone business in 2013, but that did not improve sales. Two years later, Microsoft sold most of the Nokia phone business to the Taiwanese firm Foxconn for a mere $350 million. Google purchased Motorola in 2011 for $12.5 billion and then broke up the company and sold off its different parts. Google primarily wanted the patents and intellectual property of Motorola so that Android could thrive free of legal problems. BlackBerrys also lost market share, because when all smartphones had email, it did not have the same allure—except for those people who still insisted on a tactile keyboard or needed a BlackBerry for more secure work.

By giving away Android for free, Google came to dominate the world smartphone market, as other manufacturers could take advantage of the software. Samsung started to penetrate the market in 2010 with its Galaxy S running Android. In 2018, the Chinese company Huawei became the second-largest smartphone manufacturer in the world. The largest such company was Samsung of South Korea. The American company Apple slipped into third place, though most of its iPhones were built in Chinese or Taiwanese factories. Apple retained a steady 16 percent of the smartphone market in 2020. Samsung and Huawei, as well as Vivo and Xiaomi from China, accounted for 53 percent of the rest of the market. Those four companies and many smaller ones, all running Android software, essentially accounted for more than 80 percent of the market.

Between 2012 and 2022, smartphones advanced more incrementally. They added larger screens (some so large that they were called "phablets"), better cameras (outpacing digital cameras to the extent that smartphones had a serious impact on that market), folding screens, and other changes that were incremental rather than revolutionary. Smartwatches beginning

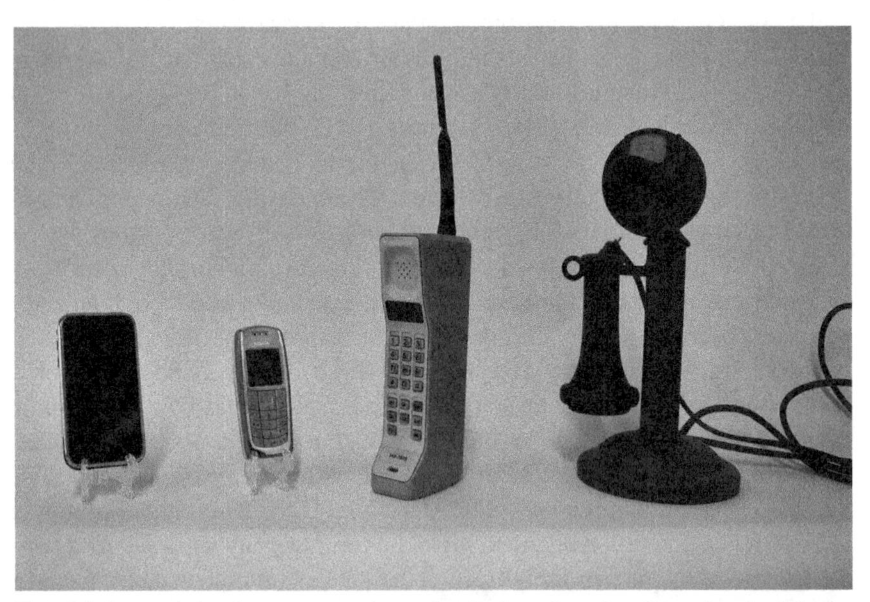

From left to right, in reverse chronological order, the first Apple iPhone (2007), the Nokia 3120 (2003), the Motorola DynaTAC "brick" cell phone (1983), and the Western Electric Operator's candlestick phone (ca. 1920). (Image Courtesy of Stella Ferro and the Digital History Archive at Weber State University)

in 2015, didn't displace smartphones either; smartphone companies still sold more than 1.5 billion units in 2019.

## TABLET COMPUTERS

While personal digital devices began appearing in people's pockets in the 1990s, other than the form-oriented GRiDPad, a larger-scale, usable tablet-style personal device did not really take off until the later 2000s. Microsoft announced the Tablet PC in 2001 and released it in 2003. It toed the line between computer and tablet with a keyboard that folded underneath the pen-enabled screen and handwriting and voice recognition. Despite the disadvantages of a tablet/PC that was both and neither, this niche—later termed *convertible* or *hybrid*—had enough momentum that Microsoft continued development. Other companies joined the movement in the 2000s. Microsoft introduced the Surface in 2013, which had a thin detachable keyboard. By 2015, for the Surface 3, Microsoft had created a single operating system (Windows 8.1) that encapsulated both PC and tablet software needs.

Apple had long been interested in the idea of tablet computers. A 1983 speech by Steve Jobs described a computer shaped like a book that you carried, that was connected to databases and other computers, and that was simple to learn and use. A prototype even existed designed by Frog Design. Ten years after that speech, in 1993, Apple introduced the inch-thick Newton MessagePad, which featured a stylus applied directly to the screen and handwriting recognition. The grayscale screen, the size of a three-by-five-inch notecard, could be used to write on, and you could edit with intuitive motions, such as scratching out words. Users unfortunately had mixed results with the handwriting recognition. It was also a bit unwieldy at 1.4 pounds and needed to connect to Ethernet by wire, although having a memory expansion slot did give it some flexibility. It could also cause a phone to dial a phone number through the sound of number tones played into a phone receiver, a callout to the early phreaking days of the microcomputer pioneers. The Newton sold well at its introduction at the annual MacWorld conference, but it never became a game changer and was discontinued in 1998. Its specialized use in health care for gathering data from patients was its most persistent use case and acted as a prototype for other subsequent systems.

Apple waited to introduce its next tablet, the iPad, until early 2010, three years after the debut of the iPhone. It used most of the same operating system for ease of creation and use. The April 3, 2010, release of the product came after many years of rumors, and it sold 300,000 units in just one day (despite numerous glitches) and a million units within a month. The iPad had a sophisticated yet simple design from designer Jonathan Ive (1967–). A great deal of attention went into the sharkskin cover that allowed it to be propped up easily. It used a six-by-eight conductive screen (it would not work with gloves) and multiple finger control. The menu items were icons with a three-dimensional look. The rechargeable lithium-ion battery, combined with a nonmultitasking approach, gave extended life on a single charge of up to ten hours. This changed how people would think about portability. There were limited physical controls: a home button, off/on, and volume. The design of the iPad had been begun prior to the iPhone, with patents filed as early as 2003, but Apple had made the decision to focus on the iPhone; much of the design of the iPad effort went into the first iPhone. Some quirky elements of the iPad included not having a calculator, because Jobs did not like the stretched look of the calculator imported from iOS. By 2020, Apple had sold over 400 million copies of this popular tablet.

Android-powered tablets followed from many manufactures. Around 2013, some pundits predicted the end of the PC in the face of the advent of the tablets. While the rise of smartphones and tablets seemed to threaten desktop PCs and laptops in the mid-2010s, their power and versatility guaranteed them an important role in the digital economy. In 2008, laptop computers, having continually increased in power, outsold desktop PCs for the first time in the United States. The biggest annual sales year for tablets was 2013. Instead of losing market share, PCs became more powerful—and in the case of laptops, thinner and lighter—with the option of touchscreens that could mimic the interface of the tablet. Consumers often had both a PC and a tablet, focusing on the tablet as a means to receive information versus the PC's orientation toward productivity. Approximately 410 million PCs shipped worldwide in 2019.

## TALKING TO COMPUTERS

The first speech recognition system, or voice user interface (VUI), did not recognize actual speech but could be triggered by audio cues. Elektro, the walking, talking, and cigarette smoking robot from Westinghouse, delighted crowds at the 1939 World's Fair in New York. Elektro's origins began when Westinghouse's Roy J. Wensley (1888–1963) designed the

Televox in the 1920s, a device that changed substation switches by recognizing sound sequences. This idea utilized existing telephone lines to do automatic, not operator-controlled, switching. Wensley built the robot for publicity, and it utilized some advanced technology at the time. The operator of the robot spoke to the robot to initiate each step of its routine, and the sequence of words and pauses (three words, pause, one word, pause, two words) triggered the movement. The operator could say any words but, of course, used a sentence that made sense for the audience, like "Will you tell (pause) your (pause) story please?"

Elektro used a 78 rpm phonograph for its audio output. The early efforts in voice reproduction—especially for the voice recording Dictaphone used in business environments—conceptually led to the desire for automatically interpreting those recordings: speech to text. Attempts to recreate human speech had begun even earlier than the phonograph, when in 1773, Christian Kratzenstein (1723–1795) created air baffles to recreate human vowel sounds while working in Denmark. In the 1930s, Homer Dudley (1896–1980) of Bell Labs created the VODER (Voice Operating Demonstrator), a speech synthesizer that was also demonstrated at the 1939 World's Fair. This work helped develop an understanding of the physics of sound needed to reproduce and interpret the human voice. John Kelly (1923–1965) at Bell Labs demonstrated a vocoder synthesizer driven by an IBM 704 in 1962, even using it to "sing" a song. A demonstration inspired a visitor, Arthur C. Clarke (1917–2008), to add HAL memorably singing that song "A Bicycle Built for Two," also known as "Daisy Bell," in *2001: A Space Odyssey*. HAL's voice in the movie was actually an actor, but Americans began hearing voice synthesis. By 1982, they could own their own talking computer with the Mattel Intellivision game console.

In 1952, Bell Labs created the "Audrey" system, which was capable of recognizing single digits as spoken by an individual on which the system had been trained and had a 97 percent accuracy rate. The public got its first glimpse of a voice recognition system at the World's Fair in Seattle in 1962 with IBM's "Shoebox." It could understand another six words beyond single digits, including "plus," "minus," and "total," so users could ask the machine to do simple math by voice. Work on speech recognition continued around the world, in Japan at NEC Laboratories and Kyoto University, at RCA Laboratories and MIT's Lincoln Lab in the United States, at University College in England, and in the Soviet Union. An early product with limited scope, the VIP-100, manufactured by Threshold Technology, even found some limited use with FedEx in package sorting. Inspired by the voice work to this point, the Defense Advanced Research Projects Agency (DARPA) created the Speech Understanding Research (SUR) program in the early 1970s and sponsored Carnegie Mellon's "Harpy" system. Harpy could understand a vocabulary of approximately 1,000 words, a considerable leap in speech recognition, by using a search algorithm that limited possibilities.

The majority of the American public had their first exposure to any kind of voice interface through telephony or, later, interactive voice response (IVR). The 411 (or 1-area code-555-1212) number began with a recording that asked, "What city?" and "What listing?" Customer assistance phone systems also often began with a recording. Systems such as the Voicepac 2000, introduced in 1970 by the Periphonics Corporation, recorded a user's response (or touch-tone keypad responses) and packaged them (potentially even speeding up the caller's audio response) for the operator. Periphonics experimented with voice recognition in the mid-1990s using the Nuance voice recognition interface combined with the Unisys natural-language processing engine, but it never went into production. In 1996, BellSouth finally brought voice recognition to a broad public through telephony when it introduced the VAL (voice portal) system. Among the issues of multiple users, multiple accents, and limited grammar, telephony systems had (and still have) to deal with poor audio. What are called *fricatives*—*f* and *s*, for example—can often sound the same coming over the line. To improve interpretation, designers created a dialog that limited the likely responses to get around these limitations.

In the 1980s, speech recognition benefited from the use of the Hidden Markov Model (HMM), a predictive model working on the probability of sounds forming words retrieved from a database. HMM also had value for other input, such as handwriting recognition. The Russian scientist Ruslan L. Stratonovich (1930–1997) first explored the Markov approach (named for Russian mathematician Andrey Markov (1856–1922), who worked in probability theory) in the 1950s, and scientists recognized its value for voice. Fred Jelinek (1932–2010) at IBM focused on a system called Tangora, which was trained by a single user to do transcription. He introduced the n-gram model, which predicted the likelihood of words within a certain distance of other recognized words. At Bell Labs, the focus on multiple speakers necessitated a different approach: looking for keywords in a context-specific dialog. Starting in the 1980s, in what has been called the statistical revolution, natural-language processing began to move from hand-coded rules to machine learning. These techniques all played a role in the increasingly large vocabulary that systems could manage in the following decades. Meanwhile, the public began to see other specialized applications with very limited and context-specific vocabulary—for example, the Worlds of Wonder company's Julie Interactive Talking Doll in 1987.

The 1990s began with a voice recognition system called Dragon Dictate for $9,000, which was outside of most consumers' price range but initially utilized in the medical profession. Released on DOS, it required users to enunciate each word because of the processing power needed for recognition. In 1997, Dragon NaturallySpeaking, although still somewhat expensive at $695, proved to be a trainable program that could recognize around 100 words per minute using the continuous automatic speech recognition technique. Both of these systems needed to be trained by a user to attain a manageable level of recognition. At MIT, generalized systems such as the Jupiter system for accessing information about weather and the Pegasus system for accessing information about airline flights worked, but they still had to limit the vocabulary they managed. That would change in the twenty-first century.

In October 2011, Apple released Siri on the iPhone 4S. One of the recognizable characteristics of the Siri app in the United States was the style of speech voiced by voice actor Susan Bennett. Bennett recorded the voice in 2005 when the application was still an experimental program being developed at Stanford Research Institute (SRI), one of a number at the behest of DARPA. Designers had directed Bennett to make Siri a bit irreverent and otherworldly and had created snappy answers to certain queries (like asking the program, "What happened to HAL?" and getting the response, "I don't want to talk about it"). This made Siri more dynamic and real in people's minds, even when the program failed to understand the user's words. Interestingly, Elektro, who had wowed audiences in the 1930s, also had snappy answers.

Siri had begun at SRI as the DARPA-sponsored Cognitive Assistant that Learns and Organizes (CALO) and Vanguard projects. The projects both began in 2003 and were oriented toward creating an artificial intelligence (AI) that learned and could assist personnel with complex, ongoing, real-time, and real-world problems. SRI's Adam Cheyer (1966–), inspired and mentored by Douglas Engelbart to augment human capability, worked on both projects. Over 500 people assisted—a huge AI program during an AI lull in the computer industry. By 2005, the program could manage many things normally expected of an administrative assistant. During a demonstration, Cheyer's work inspired Motorola manager Dag Kittlaus so much that Kittlaus went to SRI and worked to seed fund a start-up utilizing the technology. In 2007, Kittlaus gathered several people together, including Cheyer; secured $8.5 million in investment capital the following year; and targeted the recently released iPhone to host the program. Inspired by the movie *2001*, the team created the tagline "HAL's back—but this time he's good."

They called the program a "Do Engine." Other robotic assistants had been tried prior to CALO. Wildfire Communication introduced a telephone-based assistant in 1994 that did not sell well. Microsoft had introduced the animated cartoon paper clip "Clippy" (officially named Office Assistant, and the paperclip was only the default shape that very few people changed) as a Microsoft Office assistant in 1997. Most users found Clippy's energetic, "It looks like you're writing a letter," prompts unhelpful and annoying. Siri found a sweet spot for a viable assistant. Siri could, for example, with voice or textual input, not only find a restaurant nearby but also know your food preference and calendar events and find a time for lunch. Siri benefited from the need for hands-free queries on a phone, the greater internet and cell bandwidth, better voice recognition, and millions of websites available in the twenty-first century.

Cheyer could not ignore linguistic concepts, but unlike most prior systems and their traditional approach to identifying parts of speech, his team contextualized the interpretation through real-world objects. They also created relationships with forty useful web services by gaining access to each website's application program interface (API). Verizon approached the company in 2009 and made a deal to have Siri run on Android phones. But then Apple offered to buy the company in 2010 for around $200 million, and Cheyer dropped the contract with Verizon, despite already having advertisements created. Some of the founders of Siri expressed dissatisfaction with the direction that Apple took with the program, as it lost many of its defining "smart" features, and they eventually left the company. Apple had to address a much larger audience and so focused on that. Potentially, because of its size, Apple also ran into trouble maintaining all the relationships with companies that Cheyer had managed, and so Apple's Siri failed to have the same level of functionality.

With Siri and CALO as models, other similar applications rolled out, such as Nuance's Nina application, released in 2012, that coders could combine with smartphone apps, and Samsung S Voice, starting in 2013. Microsoft released Cortana in 2015. Systems were released in China by Baidu and Alibaba in 2017, and companies in Germany, Japan, and Russia also created their own. Google announced Voice Search in 2011 and released a version for both Apple's iOS and Android in 2012. Project Majel (named after the *Star Trek* actress who voiced the computer system of the starship *Enterprise*) had been started at the Google X Lab and utilized Google's superior search capability. Google later released 2012's Google Now, a proactive application that made links from the user's behavior (like typical locations and calendar events) to offer suggestions for web searches. These personal assistant programs combined AI-driven machine learning with the large amounts of data accumulated about individual users. Google Now was replaced by Google Assistant in 2016, a system that attempted to be more conversational.

In 2014, Amazon joined the party with the Amazon Echo, which used a newer category of networked technology called a *smart speaker* that typically only had a voice interface. It used the Alexa intelligent personal assistant. Google competed with the Google Nest smart speaker using the Google Assistant two years later. Others followed. While users principally used the speakers to listen to music, the companies provided SDKs. Just as in the early days of the Web and the smartphone market, numerous third-party software service companies, like Witlingo, arose to write verbal user interfaces (VUIs or, on Alexa, *Alexa Skills*) for the many companies that realized the fast-growing market for these devices. By 2020, you could call up everything from a news channel to a joke of the day. You could check your bank account, create a shopping list, and shop on Amazon or your local grocery store, all by using only your voice. Seeing the value of hands-free interaction, numerous car companies also added VUIs. Linking smartphones to a vehicle via Bluetooth allowed for orally navigating music, phone calls, texts, and more. Over 24 percent of the households in the United States had some kind of smart speaker in the home in 2019, with Amazon and Google combined having sold more than 75 percent of the 147 million units.

Many issues remained with the VUI. Users had a difficult time knowing what the systems were capable of without, for example, a GUI (graphical user interface) visual menu. Voice was only one user interface, and what designers call *multimodal designs*—taking advantage of sound, visual, touch and more—continued to develop as robust interfaces. Such interfaces included data gloves, haptic devices that gave tactile feedback, virtual reality (VR) or augmented reality (AR) goggles and rooms, heads-up displays, and wearable technology like smartwatches. The Apple Watch, first introduced in 2015, with its ability to make phone calls, had finally brought Dick Tracy's watch, the inspiration for the first cell phone, close to reality, and, better than that fictional watch, it included a VUI, motion sensor, and heart monitor.

## AMAZON

Like many innovators, Jeff Bezos (1964–) was a bright child. He took advantage of gifted children school programs in Houston and Miami. His mother recalled that when he was three years old that he took apart his crib with a screwdriver because he wanted to sleep in a bed. In his valedictorian speech in high school, he argued that humanity needed to build space colonies. Always interested in science and technology, he graduated from Princeton University with a degree in electrical engineering and computer science. Such individuals, with strong mathematical backgrounds, often found the financial industry as a comfortable alterative career, and by age thirty, Bezos was a vice president at a hedge fund. He realized in 1994 that the internet was growing so fast that it was the next business frontier, and he conceived of the idea that became Amazon. Bezos originally chose the name Cadabra for his new company, as in the magical word *abracababra.* He also liked MakeItSo.com, after the regular phrase issued by Captain Picard in *Star Trek: The Next Generation* when he agreed with the proposed plan of one of his crew. Part of the reason that Bezos settled on the name Amazon was that its first letter is first in the alphabet.

Bezos and his new wife, MacKenzie, moved from New York City to Seattle, which he had chosen because it was a technology hub. Using initial funding from friends and family, Bezos hired programmers and started to build Amazon. His intent was to build an online retail operation that sold many different types of products, but he wanted to start small and establish himself in a single market; after looking at various possible markets, Bezos settled on books. Bookstores were a market ripe for disruption, selling a product that was bulky yet had low profit margins for each individual book. Books were also fungible in that each copy of an individual book was the same as every other copy of that book. There were also a lot of books available, called the *backlist* by publishers, and only the most massive bookstores could come close to stocking every single possible book. Books were also easy to ship and did not require special care.

Online bookstores already existed, but Bezos approached the building of Amazon as a technology project rather than just an online retail operation.

He also had a laser focus on the customer. The goal was to make the experience of shopping on Amazon as customer-centric as possible and to always prioritize the customer. The advantages of online retail were the ease of maintaining a large inventory, the ease of managing a high volume of orders through a website, and being able to avoid the heavy costs of maintaining a retail storefront. The main disadvantage was that every purchase had to be shipped to the customer, making shipping costs expensive for customers and a strong disincentive to completing individual sales. Another disadvantage was the lack of immediate gratification; a customer had to wait days or even weeks for the books to arrival, which was a significant handicap when considering how important impulse buying was in a consumer market.

Amazon tried to solve the gratification problem with its Buy Now button and one-day shipping, but it was quite expensive for the customer to opt for such instant gratification. Taking advantage of its location in Washington State, Amazon also effectively avoided charging sales tax for years because such taxes were assessed by state and local governments based on the location of the seller, not the buyer. While states often had a use tax to compensate for this, the tax was paid by the buyer on the honor system, and many taxpayers did not even realize that a use tax existed. They also likely did not *want* to know that the tax existed. Eventually, once Amazon was large enough, states finally forced Amazon to start charging sales taxes.

As the dot-com boom flourished, Amazon rode the wave. As a technology company, Wall Street was less interested in traditional measures of revenue flow and profitability; it was interested in the company's narrative, technology, and potential for future growth. Bezos attracted venture capital as the company quickly grew. In 1996, sales at the company reached $16 million and the number of employees was almost 150, only a third of whom worked at the company warehouse fulfilling orders. The owners of Barnes & Noble, the giant bookstore chain, met with Bezos and proposed a partnership, and when Bezos declined, they decided to make their own website. The sites competed vigorously against each other, but Barnes & Noble still thought of itself as a bookstore first.

Amazon's initial public offering of stock on Wall Street in 1997 was not a blockbuster like other dot-com companies, but it raised $54 million, enough money to keep the company well funded. Bezos emphasized growth at all costs. He gained a reputation for being demanding and difficult, a micromanager, and a constant source of ideas, yet he was disciplined and analytical in the way that he worked through those ideas. In the following three years, Amazon used the dot-com enthusiasm to raise over $2.2 billion in bond offerings, which it used to build a network of large warehouse distribution centers and to acquire other companies. During the 1999 holiday season, a time of substantial sales for both brick-and-mortar retail stores and online stores, Amazon had five warehouses in the United States and two in Europe. In 1998, the website began to sell other products, beginning with music CDs and DVD videos, before expanding even further. The dot-com bust in late 2000 was difficult for the company; it even had to close two warehouses. However, the company had both loyal customers and a regular stream of sales, so it weathered a couple of hard years; its stock dropped for twenty-one straight months. Bezos believed in his vision and pushed forward with decisions that flew in the face of traditional business logic: he ended most advertising, believing that word of mouth was more effective, and during the 2000 and 2001 holiday seasons, Amazon offered free shipping on orders over $100. This encouraged more sales but was expensive for Amazon. In 2002, the free shipping offer was made permanent, and within months, the threshold had dropped to $25. This encouraged customers to buy more items in a single order to get the free shipping.

In 2005, this practice of free shipping led to Amazon Prime, beginning in the United States, where customers paid an annual fee and received free two-day shipping on all items. Amazon later expanded the program into selected countries where the warehouse and shipping infrastructure could support it. Amazon Prime was an expensive benefit, but it increased sales so much that it was profitable in the larger sense. Amazon also started to add other benefits to Prime memberships, such as a music services, an on-demand video service, and libraries of free Kindle e-books. By 2020, Prime had 150 million members worldwide, and one-day shipping was becoming more common, especially using the Amazon-owned shipping companies that the company had created.

The last quarter of 2001 was the first time that Amazon posted a profit for its business, a mere $5 million. This again showed that the stock market treated technology companies differently by giving Amazon seven years to become profitable. Even though Amazon continued to show anemic profits, the narrative of growth benefited Amazon, and it continued to attract investment from shareholders. Amazon also grew by acquiring companies in specialized online retail spaces, such as the online site Zappos, which sold shoes and other clothing accessories, in 2009 for $1.2 billion in stock. In 2014, Amazon purchased Twitch for $970 million, a website only three years old and mainly used for livestreaming video games as they were played, similar to how fans watch their favorite sports teams play on television. In a surprise that showed Bezos's ambition, Amazon acquired a grocery store chain, Whole Foods, in 2017, adding about 400 physical stores for $13.7 billion. This allowed Amazon to expand its online grocery offerings and to continue its experiments with mixing online and physical retail operations. Bezos was as much a business innovator as a technology innovator.

At the end of 2019, Amazon was a giant technology company with almost $300 billion in annual revenue and $115 billion in profits. Because of more revenue from technology operations such as Amazon Web Services (AWS), the company was no longer bound by the tight profit margins of online sales. An amicable divorce from his wife, in which she received her share of Amazon stock, cut Bezos's share of the company to 11 percent, though she chose to let Bezos to continue to control her stock. Even with this cut, Bezos was the world's richest person in 2020, while Bill Gates was the second richest.

Oddly enough, Bezos apparently did not earn his first billion dollars with Amazon but with an early investment in Google. Despite the dot-com crash in 2000, he continued to expand his interests. Inspired by his love of science fiction, he founded a secretive, privately owned company called Blue Origin to build cheaper rockets. Over the years, the company grew considerably, though, while launching a number of people, including Bezos himself, to the edge of space by 2022, it had still not put a rocket into orbit. Elon Musk (1971–) and his SpaceX company, who we explore below, had come to dominate that market. Even so, Blue Origin was respected for its technology and still expected to later compete with SpaceX. Among his other investments, Bezos bought the *Washington Post* in 2013, believing that he could bring innovation to a venerable example of old media. Among the more interesting innovations that Bezos guided at Amazon were the Kindle and e-books and the development of the Amazon cloud.

## E-BOOKS

Electronic books, or e-books, were an obvious innovation for computers, but the road to effective e-books proved much more difficult than might be expected. The main technology problem was that human readers wanted books to be available on an easy-to-read device that was also easy to carry around, like a real book. The main business problem was that book publishers were wary of e-books. When a person bought a book, that person now owned the book and could give it to other people, but only to one person at a time. When a person bought an electronic copy of a book, what was to prevent that person from giving copies to hundreds of friends or just posting it on the Web for anyone to download? Book publishers looked at the difficulties that the music industry went through in the 1990s and early 2000s, when electronic copies of music were easy to make and sites like Napster encouraged sharing. These problems had almost destroyed the music industry.

The arrival of PCs allowed people to read books on a computer, though obtaining books usually required someone to type a book into the computer and then share it through disks or tapes. That was a lot of work, though it helped that Project Gutenberg had been founded as early as 1971—a non-profit effort to make electronic copies of older books whose copyrights had expired, meaning the books are in the public domain. The internet made it easier to share text files, but computer screens were not mobile and not ideal for sustained reading. The PalmPilot and early flip cell phones were occasionally used to read text files, but readers had to wait until the late 1990s for dedicated e-book devices.

One early product was NuvoMedia's Rocketbook, which cost $399 and offered a 5.5-inch diagonal display; there was also an additional monthly subscription of $20 to $40 a month for the e-books. Amazon had been interested in the Rocketbook but passed on a deal to either acquire or invest in the company, but Barnes & Noble invested in the company. Another product, the Softbook Reader from Softbook Press, cost between $299 and $699.95, depending on whether you purchased a $20 monthly subscription for content, and had a 9.5-inch diagonal screen. These early e-book readers were expensive, heavy, had limited battery life, could only hold a limited number of books, and had limited content libraries of books to purchase; sustained use of the screen also caused eye strain. Though other companies also introduced competing products, the two companies hoped to sell tens of thousands of readers in 1999. However, the market failed to flourish.

A key technology for e-book readers was still being developed. In the early 1970s, the Xerox Palo Alto Research Center (PARC) developed the first electronic paper, called Gyricon. Between two layers of plastic were distributed many microscopic beads, black on one half and white on the other half; electronic charges were sent through the plastic, orienting the beads to the black half to form a dark dot or the white half to form a white dot. Partially rotating the beads gave various levels of grayscale. The advantage of this electronic paper was that it was lightweight, retained the letters or images on the paper without electrical power, and only required power when changing the letters or images on the paper. It took decades for Xerox and other researchers to make this product commercially viable, especially to drive down manufacturing costs. Sony released an e-book reader called the LIBRIe using electronic paper as the screen in 2004 for the Japanese market. The display was the equivalent of 170 pixels per inch, and four AAA batteries provided enough power for 10,000 page changes. Electronic paper was also easier on the eyes than other displays.

In 2007, Amazon followed Sony's path and released the Amazon Kindle, which also used electronic paper for its screen. Amazon found that designing and selling computer hardware was a very different business from building and running a website. Amazon recognized that readers needed a handy device that they would enjoy using and that Amazon needed to build an ecosystem that provided e-books for the readers. The success of Apple with iTunes showed that creating the media hardware was not enough—you also had to build an ecosystem. Amazon pushed for deals with publishers because Bezos wanted to announce 100,000 e-books being available when the Kindle launched. Amazon already had digital copies provided by publishers of many books because of the Search Inside the Book feature, which was used to allow customers to peek inside a book and read a bit to increase the chance of a sale. This replicated what a customer was able to do with a book inside a physical bookstore.

Even though Amazon drove hard bargains with publishers, many of them joined in the launch of the Kindle. The first Kindle had a six-inch electronic paper screen and a cell modem to connect to Amazon to download content. It sold for $399. The first manufacturing run sold out in just five and a half hours, and Amazon took months to get the product back in stock. Bezos surprised the publishers by offering best sellers for $9.99 each, even if Amazon had to take a substantial loss to pay the publisher the price that it wanted. This horrified the publishers because they feared it would set a new price expectation among customers and gut the profit margin on the sale of hardcover best sellers, which was where much of the total profit for publishers came from. Amazon eventually backed off the $9.99 price after considerable struggle.

Book publishers still wanted some form of digital rights management (DRM) software protection to inhibit the ability of customers to copy their e-books and give them away to other people or post them on the internet. The Kindle did provide that feature, though as with anything digital that must be converted to a readable form to be useful, there were ways to get around the DRM. In this way, just as the music and film industries had found, DRM functioned as a deterrent and inconvenience rather than as an absolute safeguard. Amazon released further versions of the Kindle, lowering the price and improving the quality, as often happens with most electronic devices. Kindle software for PCs and smartphones included those devices in the Kindle ecosystem. The Kindle brand expanded when Amazon entered the new tablet market with the Kindle Fire in 2011, which had a traditional LCD screen; ran the Android operating system from Google, heavily modified by Amazon; and was more about selling videos, games, music, and other media, than selling books. Not all Amazon efforts were successful. In 2014, the Fire Phone, an effort to compete in the smartphone market, failed miserably, resulting in Amazon writing off $170 million in costs and inventory.

Barnes & Noble introduced its own e-book reader in 2009, the Nook, and even dedicated substantial floor space at its retail stores to promoting the reader. The bookstore company found that it could not compete with the powerful market position of Amazon. The fortunes of the Nook followed the fortunes of Barnes & Noble; the company declined in a world where Amazon controlled a substantial share of book retailing through its website and remained the dominant player in the e-book market with its Kindle products.

The modern book took over five centuries to develop the right paper, binding, fonts, and book design to maximize ease of use. Electronic books were not an exact substitute. Studies found higher memory retention of content from reading paper books as opposed to reading a book on a screen, though that effect seemed to diminish as people adjusted to reading screens. It seemed that e-book readers worked best when reading novels because they were usually read sequentially from beginning to end. Reading textbooks or reference books as e-books proved more difficult. For instance, page numbers were harder to access, and it was harder to replicate the spatial and tactile experience of the book, which helped people retain what they read. People with good reading memories often remembered where on the page and approximately where in the book was the content they were seeking to find again. The best advantages of e-books were probably the ease by which the reader could change the font size to help with tired or aging eyes; their mobility, especially for travel; and the ease of searching text with the find function, although the find function was not always as powerful as a good index—indexes being designed by humans for humans who may be searching for a concept, not just a unique word.

## THE CLOUD

As Amazon grew, it accumulated ever greater amounts of data on its users and used data mining to help it understand its users and to provide an increasingly personalized experience for those users on the website. For instance, previous purchases of products were used to recommend possible new purchases. Most of the successful online operations were learning to use data to personalize user experiences. As part of Amazon's research efforts, the company opened an office in Palo Alto in 2003 so that they could take advantage of the talent attracted to Silicon Valley. Amazon experimented by offering its own search engine, A9 (a play on the word *algorithms*), which combined results licensed from the Google search engine with results from Amazon's own data. A9 failed to gain traction with customers, and Amazon ended the search engine after four years. However, the technical teams continued to help Amazon grow its internal technology infrastructure. Amazon also started to develop ways that developers outside the company could use Amazon data to help their own companies sell products, either on Amazon or elsewhere. Amazon was now thinking of other companies as a different type of customer and not just focusing on its own online store customers.

Amazon used the technological expertise developed for its own website and data centers to create products that could be sold to other customers. This would eventually become Amazon Web Services (AWS), the company's offering of what people came to call *the cloud* (a term first coined by Google CEO Eric Schmidt in 2006). Bezos relentlessly pushed his teams to think big, making sure that their products scaled to tens of thousands of machines.

The idea of selling CPU (central processing unit) cycles had been around for a long time, especially in the sense of time-sharing systems from the 1960s and 1970s. Computer bureaus existed to provide processing time. Faster networks gave this service much more reach. Virtual machines as a technology had been around since the 1960s and were a common feature in IBM mainframes, but virtualization took a giant leap forward in the 1990s as computers became ever more powerful. The ability of software to take advantage of parallel-processing algorithms also increased. PCs became a standard computing device, and server farms were built where thousands of rack-mounted servers, each a whole PC on a single board, formed large computers that could dynamically act as many different computers. Storage area networks (SANs) were also developed, large electronic cabinets filled with hundreds of hard drives, and the SANs used fast fiber-optic cables to connect to cabinets full of rack-mounted servers. Server software grew so sophisticated that the loss of an individual rack-mounted server or the loss of an individual hard drive did not cause problems. The dead server or hard drive was just worked around by the virtual machines running on the systems.

In 2006, Amazon started to sell online storage through its Simple Storage Service (S3) and CPU cycles through its Elastic Compute Cloud (EC2). Bezos set prices so low that the company could not readily make a profit without getting more efficient. Renamed Amazon Web Services (AWS), the products grew in a sort of stealthy way. Because this version of the cloud was not very profitable, it did not immediately attract competitors; Bezos believed that high-profit goods and services were what attracted competitors. Amazon became its own AWS customer by finally converting the last of its systems over to AWS in 2010. Eventually, the rest of the computer industry noticed that new start-ups were running their systems on AWS because it was so easy and simple. Microsoft started its Azure cloud in 2010, and Google created a Compute Engine cloud in 2012.

By 2013, AWS was contributing over $3 billion a year in revenue to Amazon. The moment for cloud technologies had arrived. Amazon continued to build data centers all around the world—large, windowless, low-lying buildings located near stable sources of electricity and access to fiber-optic networks—to provide scalable computing and storage capacity. Amazon was truly a global company, with Amazon online storefronts in most nations and global availability for AWS. AWS alone reached $35 billion in revenue in 2019 and was the leading player in the rapidly growing cloud services market. A full third of all the sites on the internet ran on AWS in 2019. For example, all of Netflix ran on AWS, delivering 125 million hours a day of content to 100 million customers in 190 different countries. Ironically enough, one of Netflix's largest competitors was the Amazon online video service, but Netflix realized that what made its company unique was its customer base, its growing library of original content, and the Netflix-specific programming code on its computing systems, not the actual hardware and networks.

In 2019, Microsoft Azure cloud was second in size, with less than half in market share, compared to AWS. Google Cloud was third. In the restricted market of China, Alibaba was the dominant cloud provider. A variety of other companies, such as IBM, Salesforce, and Oracle, made up the rest of the market. Data centers running hundreds of thousands of rack-mounted PCs had superseded the idea of the mainframe. Actual mainframes still existed, although more frequently called *enterprise computers*, mainly to service the large legacy industry because it was just too expensive to rewrite all the old code.

Much of the efficiency of modern economies came from fungibility, and much like dollar bills, electricity, and books, as noted earlier, the cloud effectively started making CPU cycles, data storage, and even network capacity into fungible services. A person could readily access as many CPU cycles or as much data storage as they needed with just a few click of a mouse (with sufficient money).

# EIGHT

# Social Media

## ONLINE COMMUNITIES

As computers and networks proliferated and grew more common, they were used to create communities. Early communities were engineers and programmers who were creating the hardware and software necessary to build the technologies. Average users started to create their own communities with Bulletin Board Systems (BBSs) in the late 1970s, as discussed in chapter 6 of this book. While many of these BBSs were quite small, CompuServe, Prodigy, and America Online (AOL) grew to total millions of users. Communities formed within these services, often centered around email lists or chat rooms. The rise of the internet in the 1990s allowed these communities to grow even more numerous. USENET news groups also formed communities. Gaming communities, starting with multiuser dungeons (MUDs), proliferated to form an enormous variety of communities.

With the rise of the Web, people wanted to have their own web pages. Some internet service providers (ISPs) gave customers server space to have their own personal websites. Setting up a website could be technically challenging, so easy-to-create web services arrived. The most prominent was GeoCities.com, founded in 1995, which was initially oriented around the geographical location of the user. Sites were clumped together into neighborhoods, which quickly changed into theme-oriented neighborhoods to group people with common interests together. A web browser–based tool allowed anyone to make their own web pages and populate them with images. The result was a bewildering mishmash of poorly designed websites that were often an assault on the eyes and any sense of design. Fan and tribute sites were common, such as homages to celebrities like Michael Jackson or Princess Diana, as well as sites for obscure hobbies, like hearse collecting and beetle fighting. Online memorials were more sobering, often dedicated to individuals or whole categories of the remembered dead, such as those who had been killed by drunk drivers.

Yahoo purchased GeoCities at the height of the dot-com boom in 1999 for $3.57 billion in Yahoo stock, taking advantage of Yahoo's stock price of $335 per share. GeoCities earned revenue through advertising, though many of the individual sites did not attract many page views. In the end, there were thirty-eight million GeoCities sites, many long since abandoned by their original creators. Yahoo closed down GeoCities in 2009. Many other sites offered similar accommodations, and while GeoCities might have been thought of as a commercial failure, it remained a fond memory for many as their first website. In 2020, there are over 330,000 companies offering web hosting, led by GoDaddy with about 20 percent of the market.

## EBAY

Silicon Valley often attracted talent from around the world. A good example is Pierre Omidyar (1961–), who was born to Iranian parents in Paris, France, and grew up in the United States after his parents moved there. He was involved with a moderately successful start-up, working as a programmer, before building a new website over the Labor Day weekend in 1995 that he called AuctionWeb. The first item he sold on the new website was a broken laser pointer to a Canadian who collected, strangely enough, broken laser pointers, an example of how the reach of the internet-enabled contacts that would have been very difficult to replicate without it. There cannot be that many collectors of broken laser pointers. The common story that AuctionWeb was created to sell Pez candy dispensers is a false story that was promoted by the company's public relations department after the site proved so successful and was renamed eBay. Omidyar firmly believed in the power of the free market and wanted to bring that power to the average person. He wanted to remove the barriers to commerce that made it difficult to readily become sellers of products. AuctionWeb brought together sellers and buyers, and he charged a small fee for each sale as well as a small percentage of the sale. A year later, Omidyar brought aboard Jeff Skoll (1965–), a Canadian, who became the cofounder of eBay when AuctionWeb was converted into eBay.

Early successes of the site were based on reselling airline tickets and the Beanie Babies craze of the mid-1990s. Beanie Babies were a perfect product to sell through eBay auctions because they were collectibles and how much a person would pay for a collectible was based on their desire for the product and how high of a price the market would bear. It helped that Ty Inc., who made and marketed the Beanie Babies for initial sale, was very canny in artificially creating scarcity for its products. In May 1997, $500,000 worth of Beanie Babies were sold on AuctionWeb, 6.6 percent of all the site volume for that month. In September 1997, AuctionWeb became eBay, and the next month, the site reached 300,000 registered users.

By its nature, connecting together far-flung buyers and sellers and providing pricing feedback in markets that had not done a good job of creating such feedback before, eBay created more efficiency in such markets. The effect was to drive down prices. For example, a dealer in Disney collectibles had to create physical catalogs to mail to possible customers, a considerable cost to just doing business, and thus marked up the prices of the collectibles to compensate. The buyer often had only the catalog for information on how valuable that Disney collectible was. Once such sellers were on eBay, it was easy for customers to find Disney collectibles that they wanted for their collection, and now the buyer could easily compare prices with other sellers. This drove down prices by as much as 90 percent, but it also allowed sellers to reach out to a much larger audience and do away with the cost of mailing out catalogs.

Auctions were exciting but an inefficient way to sell most goods. They consumed too much time and too much effort, while they could also be as intoxicating as gambling. Buyers often waited until the last moment to put in a bid, hoping to win just before the action closed. eBay expanded beyond auction sales when it added a Buy It Now button, which meant that an item could just be purchased for that price; essentially turning that part of eBay into an online retail store. Individualized digital storefronts were later added for power sellers.

From early in its existence, AuctionWeb had offered message boards where buyers and sellers formed a community. A key event came when eBay introduced a way of creating seller reputation rankings, a way to overcome the trust problem of how to get a buyer to buy something from a seller whom they have never met and who could just as easily take their money and keep it. What eBay had built was a community of trust, with buyers and sellers brought together.

As eBay grew, Omidyar and Skoll believed that the site would flourish for a while but then be overwhelmed when rising internet giants such as AOL decided to offer their own customer-to-customer auctions, so eBay had a second business of commercializing the software that ran the website. While other auction sites did emerge, as well as sites that made it easy for sellers to setup and function—such as an Amazon auction site and the zShops site by Amazon.com in 1999 (later renamed Amazon Marketplace)—eBay had created a critical mass that was hard to overcome, and the site continued to flourish. On September 21, 1998, eBay made an initial public offering (IPO) of stock, and the two founders became instant billionaires. The company was given such a high valuation by the stock market because the dot-com boom was still going strong, eBay had the potential to grow much larger, and eBay was doing something that many new internet-based companies were not doing: it actually made money.

eBay banned the sale of guns and ammunition in 1999, which provoked an outcry from vocal progun advocates, but eBay had little choice in the matter. eBay could not guarantee that sellers and buyers were following the laws of their various jurisdictions, and some items were just a lawsuit waiting to happen. The sale of many other items was forbidden, including illegal drugs and body parts. The last restriction led to enormous public attention when a seller offered a "fully functional kidney for sale." The auction price had reached $5.7 million after eight days when eBay declared the auction a hoax that also violated eBay rules. Later hoaxes included efforts to sell unborn babies, children, and a group of engineers who characterized themselves as "high-priced, professionally trained cybergeeks."

eBay also bought and sold companies, often for considerable profit. Skype was a free service that offered internet-based telephone service around the world, running the calls as internet packets instead of through traditional telecom landlines. *Skyping* became a verb, meaning phoning someone through your PC, just as *googling* had become a verb to describe doing an internet search. This worked well if people wanted to Skype from PC to PC, but the product also worked by connecting people to their cell phones or to traditional landlines. Skype was founded in Europe by Niklas Zennström (1966–) and Janus Friis (1976–), a Swede and a Dane, in 2003. Within just five years, in 2008, Skype carried thirty-three billion minutes of international phone calls, helping to shatter the monopolies that telecommunications companies had held on this market and sending prices plummeting. In 2005, eBay bought Skype for $2.5 billion, though it sold much of the company four years later. The rest of Skype was finally sold to Microsoft for $8.5 billion in 2011. eBay also bought PayPal in 2002 and spun it off into a separate company in 2015 for a considerable gain in value.

## WIKIPEDIA

*Encyclopaedia Britannica* was the most respected general encyclopedia in the English-speaking world in the twentieth century. It was the gold standard for accuracy and also the most expensive general encyclopedia, though specialized encyclopedias that covered science, technology, particular historical periods, and other narrower topics could be much more expensive. A variety of other encyclopedias existed for the general market, such as *World Book Encyclopedia*, *The New Book of Knowledge*, and *Encyclopedia Americana*. Encyclopedias were a standard feature of libraries and also found in the homes of families who valued a set of books that made it so easy to quickly look up information. In a very real way, encyclopedias are what existed before the Web came to offer similar services. All the encyclopedias were edited and written by authoritative specialists, which maintained the highest standards in this area, with entries often written by eminent experts in their subjects.

When CD-ROMs became common on PCs in the mid-1980s, several encyclopedia products emerged, taking advantage of the large amount of storage available on CDs. Early editions were text only, but later versions included maps, pictures, and multimedia experiences, such as speeches or short limited-resolution videos. These products came from established encyclopedia companies, such as Grolier and Encyclopaedia Britannica, and Microsoft joined this market with its *Encarta* product in 1993. Microsoft had tried to interest Encyclopaedia Britannica in working with them, but the encyclopedia powerhouse declined. So Microsoft obtained its content by licensing the material from the *Funk and Wagnalls Encyclopedia*. By adding more multimedia material, Microsoft sought to take advantage of the increased use of higher-resolution color monitors, PCs powerful enough to run videos, and the widespread use of sound cards. Little effort was made to expand the actual textual content.

Jimmy Wales (1966–) enjoyed reading the family's *World Book Encyclopedia* when he was younger and being schooled either at home or in schools aimed at gifted students in Huntsville, Alabama. Skilled at math, he completed a college degrees in finance and went to work for the finance industry in Chicago. Wales cofounded an internet start-up called Bomis Inc. in 1996 that created listings for a variety of different products, an example of the web portal projects that had become all the rage in the late 1990s. The company was moderately successful and provided an environment where Wales could pursue a side project based on his idea of a free online encyclopedia. This idea was not unique to Wales, but he pushed forward to make it a reality.

Wales was a fan of Ayn Rand and her objectivist philosophy and enjoyed discussing her libertarian ideas on the internet. Through these discussions, he met Larry Sanger (1968–), a doctoral student in philosophy at the Ohio State University, who was skeptical of Rand's ideas. They became friends, and in 2000, Wales hired Sanger to work on a project they called Nupedia, a free internet-based "Open Content Encyclopedia." This was planned as a profit-making business, a part of Bomis, based in San Diego.

Building an encyclopedia from the ground up was an enormous task, and a key problem was how to get the best, least biased information. Academia provided the model of using experts in a field to write the articles and to have these articles reviewed by peers who critiqued the content. This peer-review process was fundamental to the practice of science and scholarship. The process was also tedious and time-consuming. Only credentialed experts, with the proper higher education background, were expected to work on the articles. A Nupedia Advisory Board supervised the process. After eighteen months of work, Nupedia had about twenty articles completed.

The salvation of this ambitious project came from an unusual place. In 1995, a programmer, Ward Cunningham (1949–), had built a type of open-source software called WikiWikiWeb. A *wiki* was a page that the user could edit and change through a web browser, though a record was kept of every change; so anyone could always track the history of the page. This was an example of *source code control*, a method that software engineers had created to keep track of every change in the source code of programs. The hallmark of a professional programmer was one who used source code control. Though the code in a wiki could be any sort of content, it was often just prose. The word *wiki* comes from the Hawaiian word for "quick," which emphasized that the whole project was designed have things happen quickly.

Wikipedia was launched in January 2001 and was designed as ancillary project to Nupedia. The goal was to allow anyone to write encyclopedia articles, recognizing that these articles would be works in progress; Wikipedia was not replacing Nupedia but would be used to feed material into the more authoritative encyclopedia. By July, there were 6,000 articles on Wikipedia; within a year, there were 20,000 articles. Two years later, there were more than 100,000 articles. Versions of Wikipedia were also launched for languages other than English. In early 2005, there were half a million articles. A year later, there were one million articles, and in 2020, there were over six million articles in English. As the Wikipedia community grew, many of the articles were written by contributors who devoted a considerable amount of time and often contributed sophisticated specialist knowledge. The growth of new articles had slowed over the years, but the quality of individual articles kept improving.

The open-source community was already flourishing and applying very specific programming expertise to create free software. A similar impulse drove people to contribute to Wikipedia. People liked to share their expertise, even if they were not paid for the act, because people liked to belong to a community and the respect that came from sharing. The communities for open-source and Wikipedia content creation overlapped, as demonstrated when two early postings on Slashdot.org in 2001 led to notable surges in new users of Wikipedia. Slashdot, a site of "news for nerds" founded in 1997, served as a common clearing site for the open-source community.

Bomis wanted to make money on Wikipedia, but Wales realized that such an effort would damage the sense of community that sustained the site. The most obvious way to earn money was through advertising dollars, and such dollars were scarce in the aftermath of the dot-com crash. In 2003, Wales closed down Nupedia and founded the Wikimedia Foundation, a nonprofit charitable organization to own and manage Wikipedia and Wikimedia generally. Bomis signed over the rights to the new foundation. Donations from corporations and individuals funded the site, its multiple server farms, and the employees' salaries. Volunteers did the vast amount of content creation and curation.

Administrators keeping track of the site logs noticed that usage of Wikipedia always increased after Google spiders had traversed the site. The spiders updated the Google indexes so that Google searches returned the latest information, which was often found at Wikipedia. The number of people contributing to Wikipedia increased after each Google spider activity. Other sites started to copy information off Wikipedia, since it was public domain, and present the information as their own, a practice called *web scraping* or *data scraping*. Wikipedia did not try to prevent this, and sophisticated web users could readily figure out the original source of the information.

What made Wikipedia so successful? Sanger believed that its success came because the site was open to any user who wanted to contribute. Editing the articles was simple and straightforward, and there was a tolerance for articles that started off as immature efforts that grew with more content and editing by the community into mature articles. A policy of neutrality helped rein in more extreme political elements. Contributors did not sign their contributions, though a user could work their way through the Talk pages and figure out which usernames made the contributions.

Wikipedia was in many ways an organized anarchy, but rules eventually emerged to constrain problem contributors. Some contributors proved to be persistent and destructive actors, more interested in obsessively pushing their own points of view or just causing chaos. The Wikipedia community strived to work with such individuals, even advocating for "WikiLove" as the basis of tolerance, but contributors who became too much of a burden were eventually removed. Wikipedia had birthed a new way of bringing people together to collaborate and of controlling the effects of vandals, and it had created a bright example of growing reputation and trust to the world.

Wikipedia found that controversial topics, both political and social in nature, were often being vandalized by trolls or other people who strongly

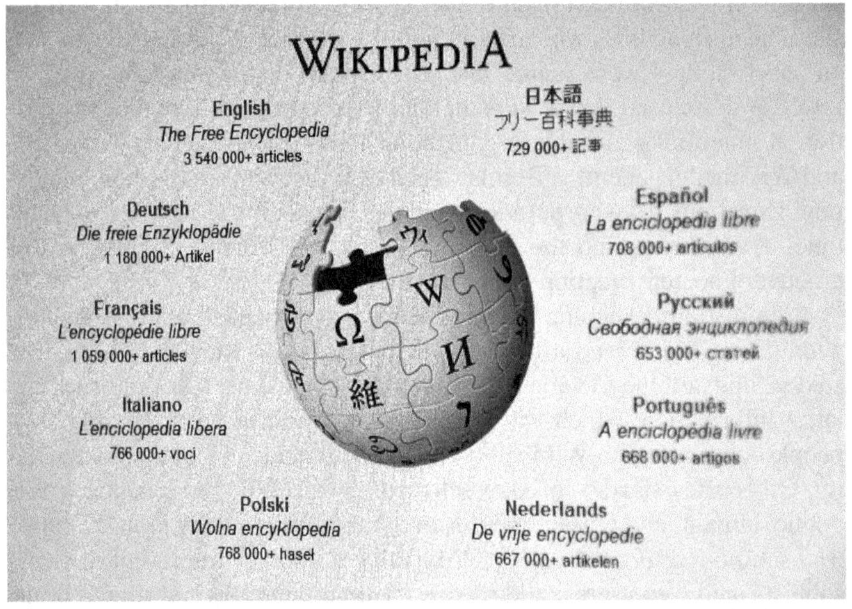


The Wikipedia web page. Despite its nonprofit status, comparatively open editing policy, and dependence on amateur editors, it became as dependable a source of information as any encyclopedia. (Nicoleta Raluca Tudor/ Dreamstime.com)

felt that their opinions were the equivalent of facts. The solution to this was to encourage contributors who disagreed to take their arguments to the Talk pages associated with each article to resolve their differences. Eventually, the solution was to lock down the disputed article and only permit trusted contributors to revise it. Trolls would have to look elsewhere to make mischief. Wikipedia also found that contacting the ISP of trolls often encouraged better behavior; being threatened by their ISP that they might lose their internet access was aggravating enough for many trolls to solve the problem. Of course, following the Wikipedia tradition, vandalism on Wikipedia was documented and described in a Wikipedia article, "Vandalism on Wikipedia," and hoaxes were documented in "Wikipedia: List of hoaxes on Wikipedia." A small number of hoaxes lasted for over ten years.

Wikipedia's goal was verified knowledge that was defined by the use of citations. Every fact or statement was expected to have one or more citations. Original research was actively discouraged, and cited sources that were available to everyone on the Web were preferred to sources that could only be found in print. Because so much of academic research was found behind paywalls, special arrangements were made to give a selected number of contributors access to such sources for free. This was to the advantage of the companies who maintained such paywalls on their content because being cited in Wikipedia encouraged other people to buy the articles. Contributors acting as Wikipedia editors actively removed articles that did not meet the generous standards of what an encyclopedia should include.

Contributors were discouraged from writing articles about themselves, even if they were prominent individuals. Contributors could even remain anonymous, though their IP address was recorded. The site exemplified the principle of crowdsourcing. While people usually did not think of Wikipedia as social media, it was social media in its own way because it fostered multiple communities and was based on user-created content. The consumers of articles formed a community, though they did not communicate with each other. Contributors formed a community of individuals committed to objectivity and verifiable knowledge and community through the Talk pages.

Wikipedia was not a traditional encyclopedia in that there was no plan. This led to complications, especially in the early stages of growth. Because contributors were self-selected, they tended to follow their interests, and the amount of material devoted to those narrow interests could be quite massive compared to other more general topics. For instance, details about individual *Star Trek* episodes could be much more detailed than a general article on sociology. A concerted effort eventually balanced out the contributions. Because the contributors were self-selected, the community of contributors was not reflective of the larger world community, and critics noted that contributors tended to be Western, white, and male, leading to some problems of bias, ethnocentric focus, and sexism. Not everyone felt welcome to make contributions and participate in the heated arguments on Talk pages.

In 2003, web traffic to Wikipedia surpassed the web traffic for Britannica.com, as the venerable encyclopedia navigated the transition from being a print product to being an internet product because sales of print encyclopedias to individuals and to schools and libraries had cratered. Even so, advocates still considered Britannica to be the gold standard because experts wrote the entries. In December 2005, *Nature* magazine published the results of an interesting investigation. Fifty articles were selected from Britannica and Wikipedia, all about science, and then sent to scientists to review for errors. Forty-two of the articles were evaluated. As a blind study, these peer reviewers did not know the source of the article that they reviewed. The study found approximately three errors in each Britannica science article and four errors in each comparable Wikipedia article. That Wikipedia had done so well, essentially a dead heat, appalled Britannica, and the company strenuously objected, even taking out a half-page advertisement in the *London Times* demanding that *Nature*'s editors retract the story.

Many scholars and experts looked at Wikipedia in horror in the early years, appalled at the lack of credentialed experts. How could anything where just anyone could contribute lead to a quality product? Wikipedia also had a dim reputation with teachers. As Wikipedia grew, its quality grew, though many of these concerns took years to fade. It was not strictly true that just anyone could edit Wikipedia. Anyone can add a new article or make an edit in many of the articles, but that new article or edit will be flagged. Then a volunteer Wikipedia editor whose responsibility includes that area will check to make sure that the contribution is not an act of vandalism.

In 2005, the first Wikimania conference was held. This annual conference provided an opportunity for contributors to come together and meet each other in person and discuss common issues. The first conference was held in Frankfurt, Germany, demonstrating how quickly Wikipedia had become an international movement. The English version of Wikipedia was only the largest of the many different Wikipedias in other languages, with each Wikipedia attracting its own community of contributors. The next conference was held at Harvard University, but since then, Wikimania has only been held in the United States one other time.

Microsoft's *Encarta* became a website as the internet grew more important, but Wikipedia kept on drawing more attention to itself. In 2009, Microsoft ended its *Encarta* encyclopedia service. The final print edition of *Encyclopaedia Britannica* was completed in 2010, and two years later, the company ceased printing anymore copies. After 244 years, the company surrendered to the inevitability of new media superseding old media. Britannica continues to be offered through the internet, and while regularly updated, it is now a niche player in an intellectual space it had once dominated. Other print encyclopedias also went online, where they struggled to find a place in the shadow of Wikipedia.

## SOCIAL MEDIA

The term *social media* first emerged in the 1990s as the popularity of the growing internet led to a proliferation of different kinds of websites. Broadly defined, many sites on the Web were social media in that they encouraged communities of users to come to the site and use their services. More strictly defined, the term came to be used to describe sites that encouraged users to come and stay and to build their own pages and content that could be used by the site to sell advertising targeted at individuals. The first social media site was SixDegrees, founded by Andrew Weinreich (~1970–). The site name referred to the idea that everyone on the planet is connected to each other within six degrees of separation. For instance, if a person has 100 friends, and each of those friends has 100 friends, then within two degrees of separation, there would be 10,000 connected people (100 multiplied by 100). Add a few more degrees of separation, with each person knowing 100 other people, then at five degrees, you have a total of 10 billion connections. At six degrees, you have a trillion possible connections. With the world having a total population of under 8 billion, everyone is theoretically connected by just five connections. Early scholarly work on this idea was done by the social psychologist Stanley Milgram (1933–1984) in the 1960s. He called this work the "small-world problem." This eventually entered common usage as "six degrees of separation." In reality, people are part of communities and social groupings that create a clustering effect, so actually reaching anyone else in the world with just six degrees of connection would be unlikely, especially if you cross geographical, class, or ethnic boundaries.

When Facebook became the dominant world social media site, there was a way to test this theory. In 2016, Facebook researchers analyzed 1.59 billion Facebook users and calculated that the average degrees of separation were a mere 3.57 degrees from each other (4.57 if you count yourself). Of course, the one in five people on the planet who were active Facebook users at the time formed a large technically literate cluster. Also complicating the relationship was that many Facebook users had hundreds or thousands of Facebook friends, many of whom they could not identify by name or their face if they actually met them in real life—although some imagined a future where an augmented interface gave people instant information about all their Facebook friends in real life via facial recognition.

SixDegrees was launched in 1997 and quickly expanded until it had about 3.5 million users two years later; then it was sold for $125 million. This was a wise move, as SixDegrees was not really successful in creating a flourishing community. A key limitation was the difficulty of getting people to upload pictures of themselves, a problem that was mostly technical in that digital cameras were in their infancy, forcing users to use scanners to upload electronic copies of physical photographs.

In 2002, a new site called Friendster, created by Jonathan Abrams, tried to realize the vision of SixDegrees and rapidly grew to millions of users and turned the word *friend* into a verb, *friended*. Digital photos were more common, and the site served in some ways as a free dating site. As the site grew, the engineering demands proved to be too much, and it went through a period of off-putting poor response to user interaction. Friendster even had the dubious distinction of becoming a Harvard Business School case study on how to not manage a technology company. In one of those missed chances, if they had known the future, a year after its founding, the owners of Friendster might not have turned down an offer by Google to buy the company for $30 million in pre-IPO Google stock. That would have eventually turned into billions of dollars. Friendster went though many permutations as it tried to thrive in a social networking market that other players came to dominate before finally disappearing in 2018.

The initial success of Friendster led to many other efforts to create social networks. One of the more successful was MySpace. Founded in 2003 by Chris DeWolfe (1965/1966–) and Thomas Anderson (1970–), MySpace passed Friendster in number of users within a year, and in less than two years, MySpace had twenty-two million unique visitors a month while Friendster had barely over a million. MySpace was focused on the music scene, as a great place for fans to meet the bands and singers they admired, but it quickly morphed into more broad content. It was so successful that two years after its founding, News Corporation bought it for $580 million. News Corporation was a mainstream media giant, the holding company for Rupert Murdoch (1931–), an Australian-born newspaper baron who had expanded into television networks, satellite services, cable stations, and movies studios, including creating the Fox family of companies. MySpace gradually faded into the background of the Web, overwhelmed by its own missteps and the success of Facebook.

## SECTION 230

For decades, federal law for the telecommunications industry had been based on the Communications Act of 1934, as modified by numerous minor legal and regulatory changes. The world of the telephone and telegraph had been superseded during that time by cable television, satellite services, and the internet. The government-sanctioned monopoly of AT&T had become just another telecommunications company. A major legal overhaul was necessary, and Congress provided the Telecommunications Act of 1996.

An important section of the new and massive set of laws was Title V, an attempt to address the concerns of people over online pornography, online violence, and other forms of obscenity. Cable television had skirted the laws that only applied to on-air services to deliver such content directly to homes. This part of the law became known as the Communications Decency Act. A year later, most of this part of the legislation was struck down by the U.S. Supreme Court, though part of the legislation survived in the form of Section 230. A major reason that Section 230 was written occurred when the online service Prodigy was successfully sued for content that a user had posted on a moderated message board. By moderating the content, Prodigy had become liable. Section 230 allowed such moderating but did not punish the provider if it failed to moderate defamatory or obscene content.

ISPs, those companies that actually carried internet traffic to people's home, wanted to not be liable for what people downloaded over their internet connections. This "safe harbor" provision also applied to user-created content, such as the content that emerging social media platforms hosted.

Advocates noted that such lack of liability sustained a vibrant explosion of free expression. Critics noted that such a safe harbor also protected bad actors and made it difficult to suppress content that most people agreed was not acceptable.

Section 230 shielded network providers and websites from liability because users had downloaded child pornography or stored child pornography on their servers. Because child pornography was illegal almost everywhere, websites still had to remove the offensive material if they were made aware of it, but they did not have to actively seek out the material for deletion. International police organizations, federal authorities, and local authorities created active programs to hunt down collections of such content, creators of such content, and the people who downloaded such content, but still the problem persisted. In essence, websites were only responsible and liable for the content that they actually created, not the content that users created. The entire business model of social media sites was based on user-created content and user-created relationships; Facebook, YouTube, Twitter, Pinterest, and the like created very little of their own content. For instance, the Yelp site is based on collecting reviews of businesses by users, and even though the whole purpose of the site is reviews, it is not liable for the reviews, though the actual reviewer could be sued for defamatory content.

As social networking sites became influential, political groups and ethnic groups used the medium to promote their points of view and, in some cases, to post hateful speech and hateful videos, even sometimes using the Web to organize violence. The problem was always that what one person viewed as cyber hatred, another person might view that same content as legitimate political discourse. This did not mean that all hateful content could be posted at all times. As private companies, the internet giants were not required to be neutral in assessing hateful content, and they could choose whether to keep user content up or to remove the content.

There were websites devoted to posting salacious or derogatory information about people. The information might or might not be true. The owners of such sites were protected by Section 230, though the posters might not be. Some sites that hosted revenge pornography, nudes or sexual photos posted by old boyfriends or former spouses, allowed people to translate their real-world anger into the cyber realm.

Activists who wanted to control cyber hatred often found their legal options blocked by Section 230. Violent extremist videos were a particular problem; they lacked a journalistic purpose, existing only to incite hate and violence. A partial solution came from the fact that the digital bits that constituted any form of digital media could be run through an algorithm to create a unique digital fingerprint, often called a *hash*. Google, Facebook, Twitter, and Microsoft chose to cooperate and share a common database of such digital fingerprints to remove pictures and videos from their sites. Such fingerprint databases were also used to identify and remove child pornography. Even with this effort, websites in the United States still had a safe harbor when it came to liability. As we will see, social media sites also struggled even with this protection in place.

## YOUTUBE

The first video on YouTube was uploaded on April 23, 2005, a mere eighteen seconds and featuring YouTube cofounder Jawed Karim (1970–) at the San Diego Zoo. Google purchased the thriving start-up for $1.65 billion a year later. While many people had no idea the site existed, Google had noticed that the site was already profitable, and it fit into Google's goals to increase more consumption of content on the Web. Google grew YouTube into the premiere destination for online videos, especially content that had been created by amateurs. By 2008, ten hours of video was being uploaded every minute, and three years later, that number had leaped to seventy-two hours per minute. Most of the content came from smaller companies or individuals, seeking exposure, fame, or fortune in the form of shared Google AdSense revenues. Five hundred hours of video was being uploaded every minute by May 2019, and by 2020, a billion hours of YouTube content was watched daily.

YouTube commercialized the site by adding advertising, but with some tweaks that made the advertising even more effective. Advertising was aimed at individual viewers based on what Google knew about that viewer's tastes, and the next video was selected using the same method, encouraging binge-watching. When an advertisement started, the user could click on a button and immediately end the advertisement and get to the video that they wanted; though if the user had watched a lot of free videos, they might actually have to sit through additional advertising. This method of opting-in and opting-out forced advertisers to make better advertisements that immediately caught and held the attention of the viewer. The old days of network television, when most of the commercials were mass media in the true sense of that term and not personalized in any way, making viewers into passive drones who watched commercials with little objection, were ending. YouTube was so successful that YouTube Red was introduced in 2015, where $9.99 a month brought YouTube to the subscriber free of advertising; the name was later changed to YouTube Premium for a higher price.

Another innovation from YouTube was to pay a small percentage of the profits from the advertisements to the content creators whose content had attracted users. Like MySpace, YouTube initially attracted a lot of musical artists. The chief problem for any musician or singer is how to reach out and find an audience. By essentially solving the distribution problem for new artists—whether they be musicians, video creators, comedians, or any other skill that could be distributed via video—YouTube created what seemed like a chaotic anarchy. Stories spread of YouTube content creators who made a living from their efforts, and a new career was born—the YouTuber.

YouTube received just criticism for its site being a hotbed of misinformation, conspiracy theories, and extremist videos. Google's goal was to run all of its services as much as possible by algorithm and minimize the direct interference of human engineers in the process. This led to examples where AI pattern-matching techniques drew conclusions that were immediately obvious as biased to a human being but not to the computer. Google corrected these problems as soon as it became aware of them, although the company did suffer some damage to its reputation. Social media had become so embedded into our global culture that people gradually came to expect social media platforms to act as more responsible social actors.

## FACEBOOK

One of the users of Friendster was a Harvard student named Mark Zuckerberg (1984–). His mother was a psychiatrist, and his father was a dentist. He was raised with upper-middle-class opportunities, and he took to computer programming early. While a nineteen-year-old student at Harvard University, he led a group of students to found Thefacebook on February 4, 2004, a social network that was originally only aimed at Harvard students. This was brilliant move because it computerized a social network that already existed, creating a digital mirror of real-world friends and relationships. Three-fourths of Harvard's students were using the site within a month. Then Zuckerberg expanded to other college campuses, adding the colleges by clumps, not all at once. Expanding gradually allowed him to scale up the site incrementally and not be overwhelmed with too many users at the beginning, as had happened to Friendster. Even so, after only two months, the site had 30,000 users.

Facebook could grow so quickly because all the pieces were available to quickly assemble a sophisticated website. By the late 1990s, it had become easy to rent physical space in server farms and place your own servers there, meaning that your company did not have to worry about physical infrastructure, electricity, network connections, or redundant systems. Renting server space in server farms became just as easy. Powerful software packages capable of running enterprise-level systems were free through the open-source movement, such as Apache for web servers and MySQL for database servers. Website creators just had to paste everything together through scripts and small programs. The term *Web 2.0* became popular later in 2004, describing both the websites that thrived based on user content, like social media, and the easy-to-use tools for making websites. Facebook was a good example of this.

After the end of the spring semester, Zuckerberg and his friends were making enough money and had attracted enough venture capital to move to a rented house in Palo Alto in Silicon Valley. Thefacebook continued to quickly grow on college campuses, and on November 30, 2004, it reached one million users. In March 2005, Zuckerberg turned down an offer by Viacom to buy the site for $75 million. Thefacebook became Facebook on September 20, 2005. A year later, Yahoo offered $1 billion in cash for the company. Again, Zuckerberg declined.

On September 26, 2006, Facebook was ready to open itself up to everyone as a potential user and exploded from eight million users to fifty million active users in a year. Children were supposed to be at least thirteen years old to join Facebook, but Facebook had no way of verifying the age of users. The new generation of children in the developed world had been raised using electronics and thought of access to the internet as natural, just as their parents had thought that telephones were a natural and expected part of their physical environment. Children lied about their age and joined the newest hip thing. More features were added, including the News Feed in 2006, the Like button in 2009, and the addition of games from the game company Zynga in 2010. These included *Farmville*, *Mafia Wars*, and *Café World*. In 2011, the standard interface for users was changed to the Timeline feature.

At the end of 2008, 70 percent of Facebook's 145 million users were from outside the United States. Facebook had now become the social media destination of choice for not just Americans but the world, with the exception of China. The Chinese government kept Facebook and other internet giants out of its large market, fostering local companies to serve the same needs. Part of this was the Chinese government's desire to help its own companies grow, but it was even more profoundly driven by a genuine fear of what social media, search engines, and unfettered communication would mean to its party's hegemony.

Facebook also grew by acquiring competing companies, such as the purchase of Instagram in 2012 for $1 billion. WhatsApp was acquired in 2014 for $19 billion, as was the virtual reality (VR) headset manufacturer Oculus in the same year. Facebook made an initial public offering (IPO) of stock in 2012, instantly turning Zuckerberg and other early investors into billionaires. The company continued massive growth on mobile devices and overseas. In 2012, Facebook started to display advertisements on its mobile applications, and within two years, this was majority of its advertising income. Mobile sources supplied 93 percent of Facebook's advertising income in 2018.

In 2013, Pew Research found that 47 percent of Americans found some of their news on Facebook, and two years later, 63 percent of Americans found some of their news from that source. Facebook itself did not generate news; the site just provided the infrastructure for news sources to push out news and the tools to target that news and provide a portion of the advertising income as incentive. News sites started tailoring their news to attract more clicks by users through more provocative titles. In the past, newspaper editors, television news editors, and other such media gatekeepers had tended to keep out the political extremes and tried to adhere to a standard of journalistic practices. That sense of professionalism began to break down, often under the guise of news purveyors describing themselves as "commentators," rather than journalists, giving them a lower standard of accuracy to aspire for. This phenomenon of low-quality news was not unique to social media and internet-based news delivery, having already become common in talk radio and talk television. The problem of false news, conspiracy theories, internet hoaxes, and extreme political partisanship bedeviled Facebook and other social media sites, and the site owners struggled in the 2010s to determine where their responsibilities lay. Were they just a platform, like a piece of paper, or were they a publisher, like a newspaper?

For the first time, on August 24, 2015, one billion unique users visited Facebook in a single day. That was almost one in every seven people on the planet. At the time, about 1.5 billion users were logging in every month. Facebook was not the result of some technological miracle breakthrough; it was just the social media company that won. That being said, Facebook and the other internet giants continued to flourish because they had built robust server infrastructures in large data centers spread around the world. Facebook also built extensive profiles of each user and their tastes, even recording mouse movements across their web pages so that they could see what users were interested in.

In April 2020, Facebook had 2.5 billion active users, reaching almost one out of every three people on the planet. WhatsApp, a Facebook subsidiary, had 2 billion active users. Instagram, another Facebook subsidiary, had 1 billion active users. Facebook Messenger had 1.3 billion active users. Google's YouTube had 2 billion active users. Snapchat had 398 million active users. Twitter had 386 million active users. Pinterest had 366 million active users. In China, Tencent's WeChat platform, founded in 2010, had 1.165 billion active users. By 2019, Tencent had successfully solved the problem of censoring in real time, as messages and images were being transmitted. This development pleased the Chinese government. TikTok, a Chinese video-sharing social network founded in 2012 that expanded to external markets five years later, had 800 million active users.

Social media had changed how the people of the world interacted. For instance, 128 million Facebook users generated almost 9 billion interactions during the 2016 presidential election. On Election Day alone, Facebook users generated 716.3 million interactions (posts, likes, views) and watched election-related videos over 460 million times. There were over 75 million tweets related to the election. Spending on digital advertising during the 2016 election reached $1.4 billion. Sadly enough, an analysis found that false news stories actually spread faster through social media than actual news stories, perhaps because people found them provocative.

## FINDING LOVE ON THE INTERNET

A specialized form of social interaction existed long before computers and networks: matchmaking. Personal advertisements for romantic partners were posted in newspapers and other locations for lonely hearts looking to expand their circle of romantic possibilities. Dating services also existed, but they were expensive and cumbersome and had a limited number of possible romantic partners. Of course, there were always bars, dances, and other socializing locations, and many cultures had formal matchmakers to facilitate marriages. An early dating service that used computers was Operation Match, incorporated as the Compatibility Research Corporation, which was created in 1965 to help students at Harvard University find dates. At that time, Harvard was male-only and in the process of becoming officially coed in a gradual merger with the female-only Radcliffe. The company expanded to other campuses and proved to be financially successful. Joan Ball (1934–) had founded a similar company in the United Kingdom a year earlier, the St. James Computer Dating Service, based on an even earlier computer dating service in Switzerland. Other companies followed.

Dating sites are not often thought of as social media sites because they are so specialized, but they also had an early start on the Web. An early obstacle for web-based internet dating was the issue of domain names. Domain names had been created in 1983 to make it easy to remember the name of an internet site. Underneath the domain names were the internet addresses. Because the federal government-owned ARPANET, it had the authority to create the Internet Assigned Numbers Authority (IANA) in 1988 as a method to assign internet addresses and domain names. In the early 1990s, a small company named Network Solutions obtained the federal contract to issue domain names for the internet. Jon Postel was still the primary individual in developing the technology for domain names. The National Science Foundation (NSF) covered the costs for Network Solutions until the number of new domains reached thousands a month. Network Solutions was then allowed to begin to charge a registration fee, $50 a year for .com, .net, or .org organizations, in 1995. There was no process to ensure that the domain names were being issued to companies or organizations that may have owned those names as trademarks or whether a person was registering a name that already belonged to another company.

Some people realized that domain names were going to become the equivalent of physical real estate, and they gobbled up domain names. Gary Kremen (1963–), a serial entrepreneur, realized that domain names were the new digital frontier and would become valuable in their own right. He registered numerous domains, and one of his greatest successes came when he registered Match.com in 1993. The site opened two years later, launching one of the earliest internet dating sites. Online dating would prove to be inexpensive compared to earlier dating services. The Match Group eventually became an international company with revenue over $1 billion per year and multiple different sites aimed at different parts of the market.

Other companies also competed in this market as online dating changed dating habits and norms around the world. A study found that by 2016, a third of new heterosexual couples in the United States were initially meeting online. Some studies found that relationships that began online were proving to be stronger, perhaps because online sites provided a larger dating pool and people could find partners with common interests more easily. Others have objected that online dating turns romance into more of a market-driven game to the detriment of human relationships and that sense of wonder that fuels romance.

Kremen had also registered Sex.com, intending for it to become a health and wellness site. A serial criminal named Stephen Cohen fraudulently tricked Network Solutions into transferring the site to him, and he made tens of millions of dollars running the site as a gateway to pornography sites. After years of effort, including litigation, Kremen managed to get the domain name returned to him, and he sold it for $14 million, whereupon the site again returned as a gateway for adult-oriented sites.

In the late 1990s, during the dot-com boom, most websites were failing to make money. The glaring exceptions were sites that sold access to pornography, politely called the *adult entertainment industry.* Adult content had long had a presence on the internet, especially through newsgroups like alt.sex and the dozens of variations, some of which distributed pornographic pictures or stories. As American laws loosened and sexual mores changed, adult content had become more common in the United States, especially in the 1960s and 1970s. The adult industry had often been an early adopter of new technologies, such as VCR tapes in the late 1970s and CD-ROMs in the 1980s. Computers made adult content creation easier, especially the invention of video cameras and digital cameras, and adult sites were pioneers in video streaming. The internet made adult content distribution even easier.

The amount of adult content on the internet proved difficult to measure. However, adult content certainly had been an important component of the internet in the past and did not flag in the twenty-first century. A 2006 study found that the most popular Google searches on cell phones were for adult content—about one out of every five searches—and this was before smartphones provided larger screens.

## GAMING REVISITED

Originally released in 1996, Nintendo's Pokémon franchise had earned a total of $90 billion by 2019, though most of that was from merchandise and visual media products. In terms of computer products, Pokémon earned $3.6 billion from mobile games, including the augmented reality (AR) game *Pokémon Go*, and $13.78 billion from console and handheld game sales. Total video game sales of Pokémon exceeded 47 million copies. Nintendo also had the second-biggest game franchise, which started in 1981, *Mario* and later *Super Mario*, totaling $15.9 billion by 2019. Activision's *Call of Duty* first-person shooter franchise, released in 2003, passed $18 billion by 2019. The Wii series of games, many of them sports or fitness oriented, taking advantage of the unique three-dimensional controller that the Wii console system introduced in 2006, had earned almost $15 billion by 2019. The venerable *Pac-Man* first debuted in arcades in 1980 and was sustained by new versions to reach $14 billion in sales by 2019. *Space Invaders*, first found in arcades in 1978, sold almost as well as *Pac-Man* the same year.

*League of Legends*, a multiplayer online battle arena game from Riot Games, earned $10 billion in its first ten years. *Fortnite*, a similar game from Epic Games, came out in 2017 and earned over $4 billion in its first two years. *League of Legends* required monthly memberships, while Fortnite joined the trend toward free-to-play (F2P) games, also referred to as *games as a service* (GaaS), where there was no monthly service fee but players could buy different skins for their avatars, different weapons, or other bling for their virtual environment. This proved remarkably successful.

In 1996, a researcher identified four types of players on multiuser role-playing games. *Killers* liked to dominate the game and dominate other players. *Achievers* liked to achieve goals in the game, like gaining levels or accumulating items. *Explorers* wanted to understand the game, not only the virtual world but also what its limitations were. *Socializers* wanted to get to know the other players in the game, using the virtual environment as a new forum to develop relationships. Of course, these were ideal types, and individual players might be various combinations of these types. Game designers deliberately optimized their games to appeal to the different types.

Psychologists argued that players were often attracted to playing video games in which they were trying to have their virtual identities reflect their ideal selves: brave, caring, rich, whatever. Trying out these traits for comfort would potentially help people bring their actual selves closer to their ideal selves.

While game technology became increasingly immersive, with more realistic imagery and greater polygon speeds to create ever more complex environments, the role of imagination was still important to create that sense of immersion. For decades, the idea of holograms or virtual reality (VR) equipment sought even more immersion. VR enthusiasm in the latter half of the 2010s failed to pan out, even though billions of dollars were spent on research and in designing equipment. An element of that may have been that a percentage of people did not handle VR interfaces well, feeling visually confused or nauseous. However, many people still believed that such interfaces would eventually come to dominate computer gaming.

## COMPUTERIZING GEOGRAPHY

Canada created a computer system that integrated geographical locations with data about those locations in the 1960s. Roger Tomlinson (1933–2014), a geographer, originated this idea, implemented it, and gave the technology its official name in a 1968 paper: geographic information systems (GIS). This revolutionized the practices of land management, city planning, and practically every application of geography. In 1993, Xerox PARC released Map Viewer, an online mapping application. MapQuest launched its efforts in 1996, delivering free map services on the Web, similar to the way that specialized GPS devices did, showing paths from one location to another with driving instructions.

Microsoft followed suit in 1998 with its experimental TerraServer, which used an imagery database five terabytes (5 TB) in size. These sites relied on satellite imagery and aerial imagery for the United States from the U.S. Geological Survey (USGS), at a resolution of each pixel being equal to 1 square meter, as well as satellite imagery taken by a Soviet military satellite, with the resolution of each pixel being equal to 1.56 square meters. The collapse of the Soviet Union had led to the data being declassified and made available for purchase. Microsoft also integrated the Encarta Virtual Globe Gazetteer with the application, connecting names with geographic locations, making the site more than a collection of images with precise coordinates. The site ran very slowly, partially because of the large amounts of data being pushed across an internet where a T1 line (1.544 Mbps) was considered fast for a corporation.

In 2001, John Hanke (1967–) cofounded a company called Keyhole, which created a program called Earthviewer. Hanke was partially inspired in developing Earthviewer by reading the dystopian cyberpunk novel *Snow Crash* (1992) by Neal Stephenson. Google acquired Keyhole three years later. Some mergers sought just the product or the intellectual rights, but the desire to hire the engineering team also motivated Google. Keyhole's technology became Google Maps and Google Earth in 2005. These products combined what MapQuest and TerraServer had offered, but it ran much faster. Google Street View followed, a project to have unmarked vans travel the United States and take pictures from every street, repeating the process every three years. Street View expanded to rest of the world, especially in urban areas. Hanke founded a new start-up within Google in 2010, Niantic Labs, with the goal of building interactive augmented games that used geographic data.

Niantic Labs released an experimental AR game in 2012 called *Ingress*, which was initially only available in a beta version. The game could only be played by people using the geolocation abilities in their smartphones, a feature that used either GPS or calculated their location with cell phone towers. Actions in the game were based on where you were physically located. Niantic Labs was spun off from Google in 2015, receiving more capital from Nintendo. They took the geolocation data that they had accumulated for *Ingress* and created an AR game called *Pokémon Go.* This game relied on the gaming mechanics and mythology of the world's most popular gaming franchise, but now players could play by watching their smartphones and interacting with an augmented reality (AR). *Pokémon Go* took advantage of the fact that a majority of people had smartphones, even teenagers and younger children.

Released first in Japan, followed by Australia and New Zealand and then the United States, this free game took nations by storm in 2016. Children, teens, and adults who were playing the game (many who had grown up playing Pokémon games) could readily be identified because they were walking around staring at their smartphones and moving them around, using the phones' cameras to see the real world with the game world overlaid in an attempt to capture Pokémon in the game. The gamers also clustered around public locations that were designated as PokéStops or Pokémon Gyms in the game. Players could purchase in-game items using real-world money, leading to a profitable revenue stream that generated over $1 billion before the end of the year.

## CRUNCHING DATA

Mathematicians in the seventeenth and eighteenth centuries developed a new field of knowledge—statistics and probability theory. William Playfair (1759–1823), a Scottish engineer, took these mathematical ideas further by creating ways of visualizing such information, including inventing the bar chart and the pie chart. Statistics promised a new way of understanding both nature and society; however, collecting the data and managing it was difficult, and processing the data by hand was tedious. Early data processing machines, such as the Hollerith tabulating machine, were a godsend for statisticians. The arrival of electronic computers added more tools. The invention of relational database technology at IBM in the 1970s by Edgar F. Codd (1923–2003) provided a new tool to organize data and code relationships within that data. In 1997, a prominent statistician, C. F. Jeff Wu (1949–), gave a public lecture, "Statistics = Data Science?" in which he argued that *statistics* should be called *data science. Data mining* became a term used in the 1990s to describe using data found in databases and finding new uses for the data and new ways to view it and extract meaning.

Some of this data took enormous amounts of processing to turn it into something useful. In the field of astronomy, radio telescopes recorded enormous amounts of data, the noise of the skies, and many scientists hoped to find hidden among all that clutter the siren song of alien civilizations trying to contact us. Processing that much data required enormous computing resources. In 1999, Berkeley SETI Research Center released SETI@home, a program that ran in the background on home personal computers (PC). The program was designed to use extra CPU cycles, especially when the computer was in screen saver mode, when the computer was just waiting for the user to come back and do something. This was an example of massive distributed processing, which worked because the problems could be worked as parallel processes, as well as an example of sharing. Soon millions of people around the world were donating their extra CPU cycles. Within just months, SETI@home was running twenty-five trillion calculations per second, equivalent to twice the speed of the most powerful contemporary supercomputer. Two decades later, SETI@home was suspended, having found no aliens and only making a small dent in the massive amount of collected data.

The science of climate change, or global warming, was heavily based on creating complex models of Earth's atmosphere. Other areas of science were also transformed by computerized data collection and the creation massive data models. While the usefulness of crunching data in science had an obvious positive effect—more knowledge and knowledge that we would be unable to acquire without computers—the ability to gather and process large amounts of data, crunching it all together, led to a new business model that came to be called *surveillance capitalism* by critics.

## SURVEILLANCE CAPITALISM

Facebook and the other internet giants flourished because of what became known as *surveillance capitalism.* Companies provided free services on the internet—search, email, messaging, social networks, or cloud storage—in return for carefully watching users and using that knowledge about each user to deliver targeted advertising. In this form of capitalism, the product that is being created is the users of the site and their personal information, and the real customers are the advertising firms who are selling advertisements to the web companies to display to those targeted users. This economic structure has also been described as parasitic and an infrastructure that could easily be used by authoritarian governments to suppress human rights and liberties.

An example of a major website that did not use surveillance capitalism was craigslist.org. In 1995, Craig Newmark (1952–), a programmer in San Francisco, started an email list of local events. The list kept growing and became a website, and because it was one of the first to offer free classified advertisements, it became the dominant player in this market. Newmark chose to not charge for users to use the site except for a few select categories, such as job listings and other services. Eventually, craigslist had to confront the problem of illegal items or services being offered, so a list of prohibited listings was created. A serious issue emerged around offering sexual services, effectively prostitution, which might or might not be illegal in local jurisdictions. Even in 2020, craigslist remained bare bones in its presentation, using text only and basic HTML, eschewing all the fancier technologies that had changed the look of the rest of the Web. Despite not trying very hard to make money, the privately held craigslist was apparently making over a billion dollars a year in revenue in 2018. An economic study found that from 2000 to 2007, craigslist saved classified ad buyers $5.4 billion.

Old media struggled to find a place in the new world of internet-enabled social media. Newspapers were particularly vulnerable because they suffered from high distribution costs in that they had to print daily newspapers and deliver a copy to each of their customers. This was a lot of physical work. A major source of income for newspapers was classified advertisements, and craigslist.org and other such sites gutted that revenue source. Obituaries even moved online. In the 2000s, newspapers started to die with such frequency that a morbid website, Newspaper Death Watch, cataloged the dying information sources.

The decline of traditional media had other important consequences. Traditional mass media was based on control of the distribution of music, films, television, books, magazines, and news. Music was delivered via compact discs (CDs) sold in stores, films were seen in theaters, a few television networks had networks of stations across the country, authors contracted with publishers who had long-term relationships with bookstores, magazines were distributed through large distribution companies to magazine racks around the nation, and news was controlled by large newspapers and large television networks. Large companies tended to dominate because they controlled how the product was delivered to customers. These distribution systems were expensive to create and maintain, and such companies tended to moderate their political or social stances to maximize the number of possible customers.

The internet changed the media landscape because now the distribution of all those media products and news was essentially free and uncontrolled. The problem became getting anyone to pay attention to your creation, not getting distribution. A television series that drew five million viewers a week in 2020 was a success yet would have been instantly canceled back in the days of the three big commercial networks (ABC, CBS, and NBC). With increased variety in channels, however, came the diminishment of the gatekeeping role that big media companies used to play. The result could be viewed as a more democratic world, but also more anarchical, as conspiracy theories and extreme political points of view also had their day in the sun. Many worried that the divide and conquer media strategy had led to ever more isolated, often virtual, communities within that society with their own sets of facts and worldviews.

## PRIVACY ON THE INTERNET

The rise of surveillance capitalism raised the issue of personal privacy on the internet. In 2020, people all over the world were often wary of governments or corporations knowing a lot about them as individuals. Some of the knowledge would be necessary for governments or corporations to do their jobs but even some of that knowledge seemed excessive to privacy advocates. As a general rule, Americans worried most about government intrusion and had created some of the world's strongest laws to prevent the government from intruding without just cause. From a legal perspective, these restrictions were grounded in the U.S. Constitution's restriction against unlawful "search and seizure." While the original intent of the Constitution was to compel judicial authorities to obtain search warrants before entering a home and seizing materials, that protection was expanded to computers and digital data. That protection was also gradually being extended to protect data about individuals held by other companies, like Google or Facebook, although Americans tended to worry less about the data that commercial companies accumulated about them. American laws to regulate such data accumulation were also weak and poorly enforced.

Europeans tended to have weaker laws preventing government intrusion, especially the accumulation of data about individuals or seizing data from individuals. Europeans, however, tended to worry more about corporations accumulating data, and they created some of the strongest laws in the world to constrict corporate-based data accumulation. The complex General Data Protection Regulation (GDPR) was issued by the European Union in 2016 and implemented after a two-year delay to allow individual nations and companies to prepare. While the law only directly applied within Europe, it reached beyond those boundaries by arguing that customers who are in Europe are protected even if the data were held outside of Europe. In parts of Europe, including Germany, an unusual practice called the *right to be forgotten* was created, where a person could obtain a court order to have negative personal information about them removed from public sources of information. In the most unusual cases, a convicted criminal might have information about their conviction removed after they have served their sentence. In practice, this right even forced newspapers to edit their physical archives to remove articles that contain that information. The right to be forgotten encapsulates the conflicting issues of whether a person had a right to totally control their own personal data, even if the data about that person served the public interest.

Complicating privacy issues around the world was the reality that most of the large internet corporations that were accumulating large troves of personal data were large American corporations like Google, Facebook, and other social media companies. Few competing companies existed in Europe. China excluded most foreign internet companies from operating in its borders and built its own competing companies for internet searches and social media that followed government policies. Chinese corporations and foreign corporations in China were expected to follow government guidance on all issues, not just privacy issues. Though companies in the United States regularly stood up to the government and took the government to court, with a similar state of affairs also seen in Europe, such examples did not exist in China.

The California Consumer Privacy Act (CCPA) was adopted in 2018 and had some similarities to the GDPR. Because many internet giants were based in California, this state law had substantial potential impact, though lobbying by those same internet giants was effective in creating a law that was substantially weaker than the GDPR. Even with the weaker CCPA that they had lobbied for in force, the internet giants started adopting policies that conformed to the GDPR because it was easier to have a single set of rules than building separate systems for separate nations.

Some people chose to take control of their own personal data through three mechanisms. First, they refused to use the large internet companies in ways that allowed those companies to keep records of them. This means that they did not use free email services, purchased items on the internet only from sites that do not keep records, and did not use other free services that rely on surveillance capitalism to make revenue. Second, they also refused to use cloud services and chose to encrypt their personal data on their own computing devices with digital privacy products using strong encryption based on public-key technologies. Such encryption was potentially inconvenient but very effective, with the rule that the larger the key, the harder it would be for government organizations like the Federal Bureau of Investigation (FBI) or the National Security Agency (NSA) to break the encryption. Third, they only used the internet through the use of encrypted browsers and anonymizing networks, such as the Tor network. Such measures slowed down their web surfing but were effective at disguising their web usage.

The Tor network was created to run on top of the regular internet, but with all traffic encrypted; ironically, the Tor software and technology were developed in 2002 by the U.S. Naval Research Laboratory (NRL) as open source, with the goal of helping political dissidents around the world hide from their governments. The Tor Project, still mostly funded by the federal government, became a nonprofit organization in Massachusetts.

## GROWTH OF THE WEB

The Web grew rapidly after Tim Berners-Lee's first page in 1991. In June 2019, the top 100 websites had 206 billion annual visits in total. This was not the measure of page hits, a popular measure early in the growth of the Web. *Hits* indicated how many files were downloaded. An individual web page could be composed of a single file or dozens of files being combined from dozens of different web servers. Each graphic, picture, sometimes even individual pixels, could be a separate file. *Visits* indicated unique visitors. The top website, with over 60 billion annual visits, was Google, with YouTube in second place at over 24 billion visits. By this point, Google was technically owned by a parent company called Alphabet, as was YouTube. Facebook came in third with almost 20 billion visits. The Chinese powerhouse Baidu had over 9 billion visitors, while Wikipedia had half as many. Two of the top ten sites in web traffic were pornography sites.

In early 2019, media analysts reported that every minute around the world, one million users logged into Facebook, 188 million emails were sent, 18.1 million texts were sent, 3.8 million searches were made on Google, 4.5 million YouTube videos were viewed, 694,444 hours of Netflix were watched, 1 million views occurred on Twitch, and almost $1 million were spent online.

Nations with repressive governments experienced the expansion of the internet differently. Ever since China joined the internet in the 1990s, access to the internet had been controlled by China's "Great Firewall," a play on words for the Great Wall of China that the ancient Chinese built to protect their nation from barbarian invasions from the north. This was not a single system but a consistent government policy of control that was implemented both from a technical perspective and through social control and propaganda efforts. Much of what the rest of the world could see on the internet was invisible to the Chinese. Comparable systems were created in many Middle Eastern nations and other nations with authoritarian or totalitarian governments.

Our narrative primarily concentrates on the United States because throughout most of the history of computers and networking, the United States was the nation with the most computers and the most influential computer companies. That began to change in the early 2000s, and that change accelerated in the following years with the rise of a large internal Chinese market dominated by Chinese computer companies. China effectively created a closed market, partially because of the language barrier but even more so because of Chinese government policies designed to protect the market from external competition. One of these large companies was Alibaba, founded by Jack Ma (1964–) and Joe Tsai (1964–). Alibaba was comparable to Amazon and became China's most valuable company in stock valuation. Baidu, founded in 2000, owned a large search engine. Weibo was like Twitter. WeChat was similar to WhatsApp. Both Baidu and Weibo were owned by Tencent Holdings, which was also the world's largest video game company, with extensive holdings around the world, such as a minority holding in Epic Games. Internet-based innovation was also thriving in China. As an example, in 2019, the delivery app Meituan grew to employ 600,000 delivery drivers on motorcycles making 400 million deliveries a year in 2,800 cities.

The internet had arrived.

# NINE

# Computers Everywhere

## UBIQUITOUS COMPUTING

By the end of the second decade of the twenty-first century, computers had become ubiquitous, embedded in so many places and things that we were surrounded by computers. By 2020, people used watches, televisions, automobiles, phones, refrigerators, and so much else from daily life that were all equipped with embedded microprocessors and autonomous sensors and most likely connected via multiple networks and using distant servers storing data, documents, pictures, and more on their health, work, and personal lives. We even became oblivious to these computers; formally "dumb" machines like blenders and washers were now computerized and networked. This was called the Internet of Things (IoT).

Embedded computing completely transformed many industries and daily life. Digital cameras became pervasive in the early 2000s, making such cameras so cheap that cameras were now found everywhere. A typical person was on dozens of cameras every day. One field changed by computers was law enforcement. Officers started to wear body cameras. Surveillance cameras in stores, in parking lots, and on streets helped to solve crimes. Smartphones kept track of their users' movements and betrayed them to the police. DNA tests confirmed the identity of criminals. DNA testing, which began in the 1980s, grew ever cheaper as the computer-driven instruments followed Moore's Law. Millions of people donated their DNA to genealogical services or health sites like 23andme.com, founded in 2006, to find out who their relatives were and what their genes told them about the probabilities of certain medical disorders. In 2018, the Golden State Killer, a serial murderer and rapist in California in the 1970s and 1980s who had stopped his crimes, was arrested because biological samples from those cold cases were used to trace the genealogy of the criminal. He had not submitted his own DNA, but distant relatives had done so, allowing forensic genetic genealogy to find him. He was not the first caught with this new technique. Some American pundits argued that if DNA genealogical testing continued to remain popular in the United States, and the data was made available to law enforcement, that no crime where DNA evidence existed would remain unsolved. A major reason that ubiquitous and embedded computers were so powerful came from being networked to each other and the rest of the world.

## NETWORKING THE WORLD

Many of the basics for networking had already been laid in the two decades before the twenty-first century. TCP/IP had become the dominant set of protocols around the world. Optical fiber networks were used for fast long-distance communication, but the "last mile" problem vexed telecommunications companies in the 1990s and early 2000s. Running an optical fiber cable in a city made economic sense because customers were so close to each other, but executives at telecommunications companies worried that running such cables to every home would be too expensive for the anticipated demand or that if they spent all that up-front capital money to run optical fiber cable to every house that they would be superseded by advances in wireless connections and be stuck with an obsolete infrastructure. Instead, most internet access occurred through existing telephone wires or through coaxial cable that had been laid for cable television and could be dual-purposed to also run internet traffic.

Satellite providers also began to offer internet connections, with the downstream traffic coming from the satellite and the upstream traffic (usually much lower in volume) flowing over telephone lines. Satellite services worked for streaming movies and surfing the Web but had too much latency to handle interactive games because it took too long for an internet packet to flow up to a satellite in geosynchronous orbit (22,236 miles above the earth) and back down and into the main internet. Cell phone service gradually increased its ability to handle large amounts of data and eventually also offered internet access for homes.

The fastest networks were always those based on fiber optics. That story began with Charles K. Kao (1933–2018), born in Shanghai, China, who fled communism with his parents in 1948 and was educated at the University of London as an electrical engineer. In 1965, based on work at Standard Telecommunication Laboratories in Harlow, United Kingdom, Kao made the breakthrough that led to optical fiber cables, glass wires built from sand, where data could be transmitted as light pulses. He later shared the 2009 Nobel Prize in Physics for this work. The two other physicists, Willard S. Boyle (1924–2011) and George E. Smith (1930–), who shared that Nobel Prize received it for relatedly inventing the charge-coupled device (CCD) in 1969, the basis for digital cameras.

Over the next two decades, fiber-optic technology continued to mature and started to make a major difference in networking. If you were using copper wires and wanted more capacity, you had to add more copper wires, while optical fiber cables were limited by the speed of the electronic equipment at each end of the cable. The faster that equipment got, as Moore's Law exercised its logic, the more data that could be pushed through the same optical fiber cable. Perhaps the only flaw was that the optical fiber cables were easy to crack by bending the cable back and forth, though the same thing could be done to copper cables, though requiring a lot more effort. The dot-com boom of the 1990s was accompanied by a parallel telecom boom, which was related in that both booms fed off each other. Both booms also crashed, leaving behind shocked investors but a twin set of gifts to all of humanity.

The telecommunications industry worldwide had always been heavily regulated by national governments. This regulation was needed because older telephone companies needed legal right of way to string their telephone wires. In return, they were required to provide telephone service for even far-flung locations. Radio and cell communications required having electromagnetic bandwidth reserved for them so that their equipment could work free of interference. Nation-states assumed regulatory control of their airwaves early on when radio showed that it required coordinated access to radio frequencies. In the United States, the telecommunications industry craved the freedom to merge with each other and to enter new markets, so Congress granted them the Telecommunications Act of 1996. A frenzy of financial activity and building of more cell networks, telephone networks, and undersea optical fiber cables ensued.

Telecommunications was a very capital-intensive industry. In the late 1990s, the money flowed, the networks were built, but the customers did not come. High-flying telecommunications companies crashed, losing a total of $2 trillion in value. WorldCom, formerly MCI, entered the biggest bankruptcy in American history up until that time. Other companies collapsed, the industry consolidated into fewer companies, and companies with money bought up the new optical fiber networks for pennies on the dollar. The enormous glut of unused network capacity—called "dark fiber" because the optical fiber lines that had been laid all around the world were not twinkling with light impulses carrying data—was eventually used in the next decade as the internet continued to grow around the world.

*Wireless technology* was literally communications technology based on using some form of radio waves to transmit data. The opposite was *wired technology*, when a physical connection must exist. When people used the word *wireless*, they usually meant the specific form of wireless technology that was used for wireless connections within local area networks (LANs). The industrial standard IEEE 802.11, first released in 1997 and upgraded regularly after, provided ever greater data rates as the underlying technology improved. This type of networking became so pervasive that it was the preferred form of networking for the IoT, for laptops, for homes, and for smartphones. Early forms of wireless networking protocols ran at 1 or 2 megabits per second (Mbps). Later protocols, like 802.11n, supported data rates up to 600 megabits. Other standards had also broken free of the constraints of LANs, measured in hundreds of feet, and now could transmit over distances that used to mean wide area networking (WAN), measured in miles, though at lower speeds. As with all wireless technologies, they only ran on prescribed radio bands, and this bandwidth had to be reserved for that purpose by every nation's telecommunications authority, which meant that creating new protocols was not just about developing the technology but also coordinating to create new government regulations around the world. In a sense, the 802.x wireless technologies competed with the cell phone technologies that were created a quarter of a century earlier.

Cell phones were also important for other reasons. Early telephones relied on copper wires, and cities in the developed world were gradually wired during the twentieth century by copper wire strung from telephone pole to telephone pole or found in cables laid through tunnels under city streets. This infrastructure was expensive and took decades to deploy. As we've seen, for countries which did not have extensive land telephone lines already laid out, cell phones were an exciting alternative. Cell towers cost less than laying phone lines to every home, and customers were able to gain widespread coverage by leaping what had looked like a required stage of technological development.

While optical fiber cables bound the whole world together, networks based on copper wiring, usually in the form of twisted-pair cables, bound the computers in companies and homes together. Cell phones provided mobile connectivity, and Wi-Fi wirelessly connected inside offices and homes. But a more intimate form of networking was also needed. The world needed a technology to allow electronic devices to communicate via low-power radio for up to ten feet.

In 1999, a new technology debuted that was named after a Viking king. Harald Bluetooth, king of Denmark and for a time king of Norway, ruled in the tenth century and strove to unite the Danes. In the 1990s, the American microchip company Intel, the Finnish cell phone company Nokia, and the Swedish networking and telecommunications company Ericsson had each been working on short-range wireless technologies. Engineers from the three companies met in Sweden in 1996 to create a Special Interest Group that would create common technical standards so that all their products would work with each other. Such collaborative efforts were an important reason why electronics were compatible with each other. The Intel engineer, Jim Kardach, was a history fan and proposed that they temporarily call the project Bluetooth while the marketing people came up with a proper name. The marketing people never came up with a better name. The Bluetooth symbol that people often see on their phones is actually two Norse runes combined, standing for *H* and for *B*, after the initials of the king.

The first Bluetooth product came out in 1999, and in the following years, Bluetooth became the preferred way to have cell phones or smartphones communicate with each other and with other devices, such as with wireless headsets. Bluetooth also fulfilled one of its original intentions in replacing printer cables between PCs and printers or scanners. Earlier efforts to achieve the same effect used infrared signals, which required line of sight to work, while Bluetooth used low-power radio waves. The range of the Bluetooth was designed to be kept under ten feet, creating a new term, personal area networks (PANs). Some versions of Bluetooth had longer ranges, and by using a concentrating device, Bluetooth could potentially transmit over more than a mile, although it was not normally used that way.

Networking thrived because of the common practice of creating industry groups to define standards so that products from different companies were interoperable. This allowed networking companies to compete by building compatible products and offering better services. On the software side, such cooperation was much slower to emerge. There were standards organizations that worked with software—such as the American National Standards Institute (ANSI), the International Organization for Standardization (ISO), and the internet's Request for Comments (RFC) process—but often a specific software product became the industry standard through dominance. Other companies had to reverse engineer the dominant software to make their own products interoperable.

A good example of this process was found in word processing software that emerged in the 1970s. All word processors were capable of creating simple text files, based on the ASCII standard, but each company created individual file formats that allowed richer content, such as defined margins, fonts, graphics, and the like. These special files could only be opened and manipulated by that particular word processing package. For instance, WordPerfect had its own proprietary formats. A company like Microsoft, who wanted to compete using Microsoft Word in a market dominated by WordPerfect, made sure that Word could open, edit, and save in WordPerfect formats. After Word became the dominant word processing package in the world, Microsoft stopped supporting the WordPerfect file formats. Microsoft's Word continued to dominate because of its market position, and other word processing programs had to cater to the de facto Microsoft standard despite existing open standard file formats for word processing.

While Microsoft had continued to dominate the word processing market during the first decades of the twenty-first century, many other forms of software had become more standards-based and had left proprietary technologies behind. This was partially a realization of a common dream in the 1980s that the computer industry could move to *open systems*, where hardware and software from different companies would be interoperable. This vision had been realized in many different ways, especially in how open-source programs undergirded the cloud and so many commercial and consumer products. True open systems had helped computers to become more ubiquitous and also helped innovation by allowing innovators to concentrate on developing hardware and software that was unique, rather than having to reinvent basic hardware, networking, operating systems, or the many types of software that the open-source movement had made available to all. Other sources of innovation were found in artificial intelligence (AI) and robotics.

## ARTIFICIAL INTELLIGENCE

In the early 1980s, the AI field started to expand commercially when expert systems showed promise. An expert system was created as a series of rules, with an inference engine, covering a narrow field of expertise.

The first successful commercial expert system configured orders for Digital Equipment Corporation (DEC) computer systems. By the end of the 1980s, the expense and limitations of expert systems became apparent, and the small AI boom collapsed. However, AI pattern recognition efforts eventually led to the ability to recognize handwritten patterns, though the person doing the writing had to be trained to keep their letters within certain parameters, as evidenced by the Graffiti handwriting recognition system that the PalmPilot introduced in the mid-1990s.

Over the decades, AI researchers developed a close relationship with cognitive science, the study of how humans think. One result of this relationship was *neural networks*, implemented in either hardware or software, that simulated the way that the human brain uses neurons and synapses to form patterns and to change the relationships within those patterns. Neural networks were computing-intensive and showed their promise as more powerful computers became available in the 1990s. Neural networks led to *machine learning*, which was a way to describe computer algorithms that modified themselves as they gained more experience. Neural networks were trained through machine learning to get better at what they did. *Deep learning* came from neural networks that had multiple layers and tried to achieve abilities that were closer to what human beings could do.

In 2012, Google announced that it had successfully trained a neural network to recognize cats in images. Google obtained ten million images that included cats from YouTube and fed them into a neural network running on 16,000 computer processors. At the end, the neural network could identify the cat inside a new image. This is something that a human baby could do with ease, and writing a program that explicitly defines what a cat looked like, no matter the angle of the camera shot, or the color patterns, or whether the cat has long hair or short hair, had proved impossible, except if you let the program train itself. Just as in voice user interfaces, this was another example of the ability of Google to use the data it had collected from users to train an AI.

In the late 2010s, a new application of images and neural networks emerged in the form of *deepfakes*. Using a form of neural networking called generative adversarial networks (GAN), an image or video could be modified to alter faces. This could be used to anonymize a person, perhaps to protect them from retaliation or other negative consequences, or it could be used to make someone look exactly like someone else in a compromising or embarrassing image or video. It was quickly used to mock or falsely incriminate political leaders and celebrities. This new ability caused anxiety, as people learned to doubt the evidence of their own eyes and to wonder whether someone was trying to deceive them. A picture or video might no longer be evidence of veracity.

As we noted in chapter 7, neural networks led to one of the most impressive achievements of AI, voice recognition. In fact, the second decade of the twenty-first century proved to be a new boom time for AI as the big internet companies snapped up graduate students in an effort to build AI into their products. The dream of these companies was to have AI solve the pesky problems that come when computers do not understand data within its context. It was easier for a human to readily recognize when a YouTube video was filled with hate or when a Facebook post was advocating violence, but having an AI program do the same proved difficult. In 2011, Google founded a division called Google Brain to focus on using neural networks to conduct machine learning. Such efforts were common ten years later at all big computer companies and at companies who wanted to realize the promise of mining big data.

Early AI pioneers tried to create programs that won at chess, drawn to an intellectual problem with well-defined rules that was a mark of high intelligence in humans. It took decades to develop a chess program that was better than the best human chess player. In 1996, world chess champion Garry Kasparov (1963–) defeated the IBM supercomputer Deep Blue in a tournament. Deep Blue then defeated Kasparov in a subsequent tournament a year later. Kasparov complained that Big Blue had been programmed to specifically defeat him, though all grand masters train to defeat specific opponents. IBM diverted the supercomputer to other projects and refused a rematch. Deep Blue was based on specialized microchips that allowed it to engage in a brute force method to play chess, analyzing hundreds of millions of moves per second.

Building on its success, IBM created Watson to take on the television quiz show *Jeopardy!* and in 2011 beat human champions Brad Rutter and Ken Jennings. Watson was able to not only search a database for answers but understand the context of the questions in natural language. In 2016, Lee Se-dol, a South Korean Go player who ranked first in the world, was beaten four games to one by an AI system built by DeepMind called AlphaGo. Go is much more complex than chess. Yet, AlphaGo was given the rules of Go, and it taught itself how to play by examining the moves in more than 100,000 matches. Google purchased DeepMind and a successor AI system, Agent57, which later taught itself to play 57 classic Atari games in 2020 and proved it could play the games better than humans.

Many commercial computer games and video games had rudimentary AI algorithms to provide an artificial opponent for the human player, with decidedly mixed success at being a challenge. Many canny players found the conceptual blind spots in the game AI and ruthlessly exploited them to their advantage. This was one of the reasons why online gaming was so popular: gamers loved the challenge of competing with real people rather than learning to trick a dim-witted AI. While having an AI play games was fun, the ultimate goal of AI researchers remained recreating human-level intelligence.

The term *general AI*, also called *hard AI* or *strong AI*, described creating an AI that was equivalent to a human being in problem-solving and reasoning and was conscious of itself. This last requirement had been very difficult to achieve, especially considering the fact that understanding our own consciousness was difficult. Psychologists, neurologists, cognitive scientists, philosophers, and computer scientists all worked on this difficult problem. How would we know if other higher animals, like dogs, cats, elephants, and dolphins, are conscious, and what neurological mechanism made us self-aware?

In 1950, Alan Turing proposed a test called an *imitation game*, or *Turing test*. A person acting as the tester was isolated from an individual and a computer running a program, and the tester asked questions of each to try to figure out which one was a human and which one was a computer. The questions were transmitted through a teleprinter or intermediary so that the communications method itself would not give any clues, only the content of the answers would be relevant. The purpose of the imitation game was to see whether a computer can imitate a human. The Turing test had been very influential in the history of AI, but it later morphed into a more rigorous test to determine whether the computer was as intelligent as a human, rather than whether the computer could imitate a human. As of 2020, no computer had come close to passing a rigorous Turing test.

In 1993, Vernor Vinge (1944–), a mathematics and computer science professor who had also gained renown as a science fiction writer, wrote an essay titled "The Coming Technological Singularity: How to Survive in the Post-Human Era." He argued that the speed of development in computers would lead to superhuman intelligence, an event that he called the "singularity." Vinge expected the technological singularity to occur after 2005 but before 2030. The concept became popular in science fiction, and Ray Kurzweil (1948–), a successful computer entrepreneur, technological enthusiast, transhumanist, and optimist about the potential of achieving strong AI, adopted the term and advocated that we were within sight of achieving this goal. Other AI researchers pointed out that current AI systems were brittle, easily breaking when encountering data or situations they were not trained to handle. We had ideas of how to achieve superintelligence but no obvious path forward.

Although it was difficult for researchers to postulate what it meant to be smarter than a human, given that humans had never, that they knew, encountered a more intelligent entity, there was no reason to suppose that we were at the apex of all possible intelligences. At the very least, it was imagined, an AI smarter than a human would be smarter than the smartest person to ever live and would also be able to do many tasks in parallel by duplicating its processing power. Such a superintelligence would theoretically be able to be more capable, more creative, and more innovative than a human being.

The implications of strong AI had been examined in science fiction books and films for years. In the 1968 film *2001: A Space Odyssey*, the HAL 9000, a computer aboard a spaceship bound for Jupiter, had a pleasant male voice and seemed helpful, until it started to murder the astronauts. The same voice then sounded quite menacing. The fictional computer had been given two contradictory directives that led it to conclude that killing the astronauts was the best solution. Two years later, in the film *Colossus: The Forbin Project*, supercomputers were placed in control of both the American and Soviet nuclear arsenals. Bad things happened. In the *Terminator* and *Matrix* films, computers took over the world and waged war on humanity. These films, as well as stories, media, and games like them, represent our fears of AI.

It stood to reason that if we could create an AI with human-level intelligence, superintelligence would quickly follow, as the superintelligence would be just more of the same of whatever technique got us to AI. When people like scholar of cognitive science, Douglas Hofstadter (1945–), cofounder of Sun Microsystems, Bill Joy (1954–), Bill Gates, and Elon Musk expressed concern about an AI superintelligence, their concerns were considered by many as legitimate because of the likelihood we would almost certainly eventually create such a superintelligence. The argument for that certainty was as follows: if the human brain is a physical object and we can duplicate what it does in a physical computer, and if we continue to make computers ever more powerful, then we will eventually create a superintelligent AI. The counterargument was that consciousness may be a physical phenomenon, but perhaps it occurs on a quantum level and cannot be duplicated outside of biological systems. Another argument that people often advanced was that consciousness was a soul or spirit and cannot be duplicated. This dualistic argument—arguing that humans are a combination of body and spirit—while rejected by contemporary science, proved difficult to refute. The Google search engine, with access to the massive amount of data on Wikipedia and the World Wide Web (WWW), had the knowledge that researchers expected a superintelligence to have access to, yet none of these achievements had the consciousness or capabilities that we imagined for an AI. Some computer scientists argued that intelligence could not be created inside a computer because such an AI would be deprived of a sensory experience of the real world, so perhaps the first AI would occur in a robot rather than inside a sterile computer room.

## ROBOTICS

Robotics has been associated with AI because many of the difficult problems in both fields have been similar. Industrial robots that performed limited tasks became common in advanced factories in the 1980s, while autonomous robots that could correctly perceive the natural world through vision or other sensory means and react to that sensory data remained the dream that drove continued robotics research. Initial efforts in robotics quickly revealed that what humans or animals did naturally, with no effort, was very difficult. In 1969, a robot called Shakey at Stanford University required five minutes to move a foot, even though the room contained few obstacles. The computer processing power required by the programming of Shakey could not be carried aboard the five-foot-tall robot, so an external computer was attached to the robot by a cable. Many researchers thought that the problems of programming autonomous robots would be solved with ever more powerful computers, but that proved only partially true. Significant work at MIT's AI Lab in the 1990s developed a different bottom-up approach to robotics, creating machines that moved like insects using simple algorithms.

Researchers developed robots connected to an operator by remote controls as well as more autonomous machines. With remote control, an operator guided the robot through a controller and screen that gave the operator video and sensor feedback, such as a bomb disposal robot, where the operator was far enough away to survive if a mistake was made. One approach to improve the experience was to create better communication between operator and robot. For example, full telepresence control over the robot might include feedback similar to a virtual reality (VR) environment. Researchers also strove for more autonomy: either robots that followed simple scripts and did the same thing over and over again or more sophisticated autonomous robots that were expected to analyze their environment and make decisions.

Some of the more glamourous successes of robotics came from the American Mars missions of the National Aeronautics and Space Administration (NASA). In 1997, the Mars Pathfinder mission landed the robotic rover called Sojourner, which crawled for short distances under remote control from Earth. In 2004, two further rovers were successfully landed to search for geological evidence of water and past life, and they traveled up to 100 meters a day in a semiautonomous mode. These robots were a combination of remote control, in that the robots fulfilled instructions sent by radio from Earth, and being partially autonomous, in that because the instructions took many minutes to be sent and the results returned, the robots could do some actions on their own while waiting for the bigger decisions from Earth.

The company iRobot introduced the Roomba in 2002 after basic toy robots had been on the market for two decades. These flat-topped round devices moved around a house, vacuuming up dirt and inadvertently entertaining children and pets. Robots in the home had moved beyond toys. iRobot had been founded by roboticists from the Massachusetts Institute of Technology (MIT) in 1990, building both professional and home robots, and by 2020 had sold over thirty million home robots. Many robots for consumers, like the Roomba, were autonomous. Other consumer robots were mostly remote controlled, like drones and other toys.

A large percentage of robots were industrial robots, usually found in factories, sometimes costing millions of dollars. These types of robots revolutionized manufacturing, doing welding, assembly, and handling delicate materials, like silicon wafers for microchips or glass windshields for automobiles, and cut the need for human factory workers. Professional service robots helped surgeons operate and farmers milk cows, which are both examples of remotely controlled devices that rely on the intelligence and skills of their human operators.

Industrial robots became pervasive in manufacturing environments by 2020, even in China, whose rise as a manufacturing powerhouse was based on inexpensive human labor. Japan had led the way in both industrial and

The MakerBot 3D plastic printer. An example of a robotic system that could be found in the home. First created in 1981 by Hideo Kodama in Japan as a rapid prototyping system, 3D printers became cheap enough so that many consumers could afford them by 2009. (Oleksandr Lutsenko/Dreamstime.com)

personal robots, but in 2013, China became the world leader in having the largest base of installed industrial robots. The other three countries with large installations of industrial robots were the United States, South Korea, and Germany. The world automobile industry was strongly dependent on robots, followed by the electronics industry. On a global scale, counting the manufacturing industries in heavily industrialized countries, there were 99 robots for every 10,000 employees in 2018. Lesser developed countries had few, if any, robots. In 2018, it was estimated that there were 2.4 million industrial robots in the world.

An example of industrial robots from 2020 was the Japanese manufacturing line that assembled PlayStation 4 game consoles. It was a mere 103 feet long, and only four humans worked on the line, two at the beginning, placing bare motherboards on the moving assembly line, and two at the end, packaging each console as it rolled to the end of production. All of the other work was done by robots, including attaching wires, twisting cables into the right place, and applying tape. Every thirty seconds, a new console rolled off the line.

When fantasies of robots were imagined in science fiction, we were not enchanted by visions of industrial robots but those of artificial people driven by AI. The dream of a general-purpose robot was quite similar to the dream of a general-purpose AI, and both became popular images, such as R2-D2 and C-3PO from *Star Wars* (1977) or Data from *Star Trek: The Next Generation* (1987–1994). The TARS robot in *Interstellar* (2014) reminded us to abandon our images of robots requiring specialized legs like humans. That kind of movement—the film hired actual researchers in robotics during filming—showed the kind of innovative thinking that robot designers pursued. These were images of helpful robots. Alternatively, we have imagined darker images, such as the Terminator from the *Terminator* (1984) films and Ava in *Ex Machina* (2014), a humanlike robot who proved "her" intelligence by deceiving and escaping from her creator.

There were attempts to make general-purpose robots that look like human beings, especially in Japan, an effort to help the massive number of elderly people that needed care. One issue discovered was the "uncanny valley," also found in video games and other media. Researchers had shown that the human brain was exquisitely wired to carefully detect emotional nuance in human faces, a useful trait for creating social relationships. So when an avatar's face in a video game resembled a human too much, an anime or animated character's face was too lifelike, or a robot's face was supposed to look human, people experienced a sense of wrongness.

We also created flying robots, otherwise called *drones*. Radio-controlled (RC) airplanes were created in the first half of the twentieth century, designed as sophisticated toys and prototypes. During World War II, the U.S. military used radio-controlled drones for target practice, and a couple of secret projects tried to turn full-sized bombers into remote-controlled bombs to fly into difficult targets. While the projects were not successful, the principle of radio-controlled airplanes was firmly established. The American arsenal of drones can be traced back to a DARPA project of the early 1970s. John Stuart Foster Jr. (1922–), an enthusiast of remote-controlled airplanes as a hobby and a Pentagon executive, pushed DARPA to build two prototypes, named *Praeire* and *Calere*. These drones were remotely controlled from the ground, just like hobby radio-controlled aircraft.

The Pentagon had already built large drones to be carried by the secret SR-71 reconnaissance airplane on spying missions over China and the

Soviet Union in 1969–1971. The drone would launch from the SR-71; fly over a target, following the instructions in its navigation system; take pictures; and then return with the undeveloped film for development. The film canister was ejected from the drone for recovery at sea while the drone self-destructed, as landing was far beyond the capabilities of a drone that was not being remotely controlled. All missions apparently failed. Far more successful were the early spy satellites in the 1960s, which worked by orbiting over foreign countries, taking pictures, and then jettisoning the undeveloped film back to Earth.

What made drones particularly useful was the rise of digital cameras. Remote operators could now observe in real time what was happening to the drone. The Americans deployed an unmanned aerial vehicle (UAV) called the Predator over the conflict in the Balkans in the mid-1990s. With its large wings and lack of a pilot, the remote-controlled Predator could loiter for twenty-four hours. While the first Predators merely carried cameras, the Pentagon tested the idea of mounting Hellfire missiles on the Predators in 2001. The Hellfire had been developed to be fired from helicopters. This effort was successful. Remote operators of the Predators aimed a laser from the drone and fired the Hellfire, and the missile used its own camera to home in on where the laser was pointed.

The air force was already working on the next version of the Predator, which was given the ominous name Reaper. With a wingspan of over sixty-five feet, the Reaper could stay aloft for fourteen hours and fly as high as 50,000 feet. Its weapons load of 3,800 pounds included Hellfire missiles and GPS-guided bombs. By 2009, the air force was training more personnel as drone-joystick pilots than pilots who sat in cockpits. Many in the pilot-centric air force were not happy with an emerging world where nonpilots sitting at consoles in trailers halfway around the world, very similar to playing a sophisticated video game, were as important as men and women pilots who had spent years acquiring their flying skills and putting their lives on the line in the field. By 2020, the air force had announced research into drones that might effectively replace fighters flown by human pilots or act as wingmen for pilots in the air.

Numerous similar types of drones had been made by dozens of countries around the world by 2020. Commercial drones that were sold as toys or for peaceful purposes had been purchased by terrorist groups or insurgent groups and been adapted to be used as cheap reconnaissance assets or cheap bombs on the battlefield. All these drones were remotely controlled rather than autonomous drones. Experiments had begun on creating truly autonomous drones, and that concerned many people as an ominous escalation of military ability.

A video called *Slaughterbots* went viral in 2017 and gained over two million views on YouTube. The video portrays small drones that have cameras and connections to a database using facial recognition technology, and they also carry a small explosive charge. In the video, thousands of the robots are released by terrorists, which then proceed to track down their designated targets and kill them by diving into their skulls and setting off the explosive. The video ends with a plea by Stuart J. Russell (1962–), an AI researcher and professor of computer science at the University of California, Berkeley, to join a campaign to ban lethal autonomous weapons. The proposed method to implement this ban was a United Nations treaty. Numerous academics and experts signed an open letter for the campaign, including such luminaries as Stephen Hawking (1942–2018), Elon Musk, and Steve Wozniak. Their fear was that we were building robotic weapons that would kill without a human being involved in the decision loop. Critics argued that such concerns were overblown because such weapons would be strictly controlled by military forces and could easily be defended against. A dissimilar movie to *Slaughterbots*, showing the complexity involved in numerous humans overseeing robot behavior, in this case a potential drone strike, was 2015's *Eye in the Sky*.

One of the more obvious lessons of the history of military technology was that as new weapons were invented, those weapons would eventually find their way into the hands of smaller groups of people. This trend would allow an individual to have ever more powerful weapons, whether guns, chemical weapons, bioweapons, or even nuclear weapons. Many people expressed fear that autonomous robots that combined AI and stealth could make an individual or small group of people that much more capable of causing damage.

## DRIVERLESS CARS

In 2001, Congress required that a third of U.S. Army vehicles be driverless by 2015. Defense contractors approached this problem in a slow, incremental manner, so DARPA decided to goose the chase by announcing the DARPA Grand Challenge in 2003. Teams were invited to build autonomous vehicles for a race in the Mojave Desert over a 142-mile-long course that included switchbacks, gullies, the possibility of random animals, and other navigation obstacles. DARPA hired some professional off-road drivers to drive the course and verify that it was a challenge even for them. The first team to complete the course would win $1 million. Teams that accepted the challenge included a high school; a robotics company that had been building battle robots for shows like *Robot Wars*, where primitive robots on wheels try to destroy each other; start-ups; and numerous other teams from established companies and universities.

In all, 106 teams applied and wrote the required technical papers, and the 25 most promising were invited to qualifying rounds at the California Speedway in Fontana, where their vehicles were inspected for safety. Fifteen teams were invited to the actual race, though the actual course was not revealed until two hours before the event, when each team was given a CD with the waypoints for the course. One participant thought that the vehicles, bulging with sensors and modifications, looked like they were out of a *Mad Max* movie. Vehicles used GPS, cameras, laser rangefinders (lidar), and radar sensors. Vehicles were released onto the course one at a time, and multiple crashes quickly occurred as vehicles hit obstacles. One participant even entered a driverless motorcycle, reasoning that it would be able to go fastest over the obstacle course, and though it failed quickly, he went on to cofound Google's driverless car effort. The biggest vehicle, a fourteen-ton Oshkosh truck for Team TerraMax, got stuck between two tumbleweeds, moving back and forth because its software saw the flimsy weeds as immovable objects. Four vehicles made it more than an hour into the race, but none got farther than seven and a half miles.

This was a perfect example of failing into success. DARPA announced a second challenge for the following year, doubling the prize, and planned to keep doing the challenge every year until someone found success. Those extra contests were not needed because the second challenge produced 195 teams, and five vehicles completed the course. The winner was Stanley from Stanford University, which took under seven hours to go over the 132-mile course. Two years later, DARPA created an Urban Challenge, again for $2 million, and six out of eleven teams completed the course. Though Congress' original charge that one-third of army vehicles be driverless by 2015 was not met, the investment by DARPA was fruitful.

These contests created a community of bright, innovative people who believed in the promise of driverless technology and continued to work in the field. DARPA used the term *autonomous vehicle*, though the word *driverless* became more popular. Another participant went on to invent 3D lidar (light detection and ranging) technology. These contests were a good example of some minor funding by the federal government launching what promised to be a multibillion-dollar industry. DARPA was excited enough by these successes to create other challenges for robotics, for radios that worked in the presence of radio spectrum interference, and for fully automated computer network defenses that could resist cyberwar attacks.

Even before the DARPA Grand Challenge, various efforts to make driverless cars had been made over the years, usually based on sensors in or along the road to guide the car and some sort of centralized system that controlled all the cars on the road. The goal for truly driverless cars was to create a car that could share the road with human drivers in other cars and did not need additional help from a central system or roads that had been designed to help the driverless car. In other words, the goal was a car that basically replaced the human driver. The promise of this technology was immense: fewer accidents, reduced fatalities, increased usable time in people's lives because they could do other things while riding in a car instead of driving, and senior citizens and other physically challenged people being able to use individualized transportation even though they lacked the ability to drive. Google started its effort to build driverless cars by buying a start-up that had been spawned by the DARPA challenge. In 2016, the effort was spun off as a separate company called Waymo. The company had already driven more than two million miles on public streets by 2022 with its driverless research vehicles. Other companies created their own driverless car efforts, such as Tesla, which we explore below, and Uber, a ride providing service started in 2009 as Ubercab by Garrett Camp (1978–) that utilized computer technology to successfully compete against traditional taxi services.

The future for driverless cars looked bright, but the cars showed themselves to be very capable only in normal driving and weather conditions—much work remained. The cars got confused by rain and slick roads, snow on the road, and leaves blowing across the road. In one tragic case, the car confused the side of a truck with the sky. These wrinkles meant that a human driver had to step in and drive the car. Unfortunately, having humans step in at critical times during the most challenging situations required experience and skill, which were things not learned if the vehicle drives autonomously most of the time. This is something the airline industry recognized with its pilots back in the 1970s, and it began focusing even more on pilot training. Autonomous driving also created ethical situations that were once the realm of purely speculative fiction. For example, during an accident, should the car try to maximize the chance of passenger survival or the survival chances of pedestrians? It is the classic trolley dilemma scenario in ethical philosophy.

People often demanded certain exacting standards for how computers should work and accepted, at some level intellectually, that computers were only as reliable as their programming, but they were simultaneously emotionally intolerant of computers making mistakes. People expected humans to regularly fail and accepted that a certain percentage of people would die in automobile accidents, but they were appalled when a self-driving car made a mistake and killed someone. This reaction persisted despite people understanding that widespread use of self-driving automobiles would save many more lives compared to letting humans continue to drive.

Elon Musk pioneered many driverless car features through his company, Tesla. He is an example of an entrepreneur who began in the traditional computer industry, developing companies as part of the dot-com boom, but recognized that computers had the potential for changing other industries. His companies, SpaceX and Tesla, have applied embedded computing to rockets and electric cars, respectively.

## ELON MUSK

Born in South Africa, Elon Musk showed an early interest in programming as well as a fondness for science fiction. He also read the entire *Encyclopaedia Britannica* as a child, filling his photographic memory with facts. After attending college in Canada and the United States, earning a combined degree in economics and physics from the University of Pennsylvania, Musk became a serial entrepreneur in California. His first company, founded with his brother in 1995, was Global Link Information Network, later renamed Zip2, a site to provide internet directories for local retail companies and newspapers. Compaq purchased the company in 1999, and Musk walked away with $22 million.

Musk moved on to found an online bank named X.com. His bank merged with a competitor named Confinity, who was already building an online payment system called PayPal that would allow people to easily send and receive money with no more than an email address. The two cultures of the company struggled to mesh together, and a technical argument led to Musk being forced out of company while he was on his honeymoon in September 2000. The Confinity engineers favored open-source software, like Linux, to form the basis of their software. This was a common attitude among internet start-ups. However, the open-source movement was still maturing, and certain programmer productivity tools were still not fully developed. Musk preferred to rely on the more mature Microsoft programmer productivity tools, which worked best with Microsoft enterprise servers. Such a dispute may sound trivial, but in the heated programming wars between the open-source movement and Microsoft in the 1990s and early 2000s, such questions were treated with ideological fervor. At that moment in time, Linux servers had a well-deserved reputation for being rock solid. Microsoft servers were not as reliable or as efficient, but Microsoft programming tools were the best in the industry.

X.com became PayPal a few months later, and even though Musk was now only an adviser to the company, he continued to increase his investment share in the company. The dot-com crash happened, but PayPal continued to flourish and even rejected buyout offers until eBay offered $1.5 billion. Musk's share was $250 million. Now Musk did something even more interesting. He took his financial stake and invested in two companies that on the surface were not computer companies, but they essentially were because computers had become pervasive in almost every industry, increasing productivity and often changing how an industry's products worked.

Musk decided he wanted to build rockets, because cheaper rockets would lead to the science fiction future in space that he had dreamed about as a child. The space launch industry was a mature industry led by large aerospace companies who were comfortable with NASA, military, and commercial satellite contracts. Putting stuff into orbit still cost about $10,000 per pound. Rockets were only used once, and the space shuttle, which had been designed to be reused, had proven more expensive than planned. Musk dreamed of going to Mars, a much more difficult project than the NASA-guided Project Apollo of the 1960s to go to the moon. Space industry advocates had argued for years that if the cost of low Earth orbit (LEO) was reduced to $1,000 a pound, lots of interesting developments became feasible, such as space tourism, more satellites, and more commercial activity in space.

The space launch industry knew how to build rockets and had decades of experience doing so, so radically new rocket technology was not needed, just better ways of building rockets. Musk read books on how to build rockets and moved to Los Angeles to be close to the space industry and take advantage of the concentration of aerospace engineers in that area. On May 6, 2002, even before the sale of PayPal had closed, Musk founded

SpaceX. The company developed a new rocket engine that was highly reusable and built rockets in which the first stage would return to Earth's surface and land vertically, a trick only possible with automated computerized controls. The company concentrated its development, test, and manufacturing engineering in a single location in Hawthorne, California, a former site for manufacturing parts of the Boeing 747 jetliner. This physical concentration dramatically reduced the costs of communication and coordination between the different teams in the development and manufacturing cycle.

SpaceX relentlessly drove down costs, and by 2018, it was putting satellites into orbit for $1,200 a pound in the Falcon 9 rocket and $640 a pound in the Falcon Heavy rocket. SpaceX had taken over the commercial launch business and put over half of the satellites being launched worldwide into orbit. On May 30, 2020, the SpaceX-designed Dragon capsule flew into orbit on a Falcon 9 rocket with NASA astronauts to visit the International Space Station, the first crewed American space launch since the 2011 retirement of the space shuttle. Five months later, four more NASA astronauts flew in a SpaceX Dragon.

Musk's other new company, Tesla Motors, was an electric car company, founded in 2003. Musk became its largest investor in its initial funding round and eventually took over management of the struggling car company in 2008. Earlier electrical cars had relied on lead-acid batteries or nickel metal hydride batteries, but new batteries in the form of lithium-ion batteries were much more efficient. Earlier electric cars had developed a reputation for accelerating slowly and being unable to travel far between charges. Tesla focused on the Roadster as its first product, a premium sports car that would accelerate fast and travel a fair distance. Tesla charged a premium price to a select audience and successfully changed the public perception of electric cars as inferior to traditional cars that relied on the internal combustion of fossil fuels.

Tesla introduced a more mainstream model, the Model S, in 2012, a five-person, all-wheel drive sedan that could accelerate from zero to sixty miles per hour in just 2.4 seconds and travel 345 miles on a complete battery charge. Musk described the Model S as a "very sophisticated computer on wheels." Onboard computer control over the car's batteries and engine was vital to making the vehicle efficient. The company could send software updates to each car's computers, adding new features and refining the operation of the car, a major change in the industry and a huge benefit for solving problems and improving the car without the car coming into a mechanic's shop. In 2018, for example, the company responded to criticism of the Model 3 braking distance by sending a software update to all cars that improved the braking distance by nineteen feet.

There was also an autopilot feature that allowed the car to drive itself or to perform complex maneuvers on its own, such as parallel parking. The autopilot feature relied on rear-, side-, and forward-facing cameras; a forward-facing radar; and twelve ultrasonic sensors around the car. Tesla struggled with the issue that the autopilot feature was not truly autonomous and that drivers were supposed to maintain control and actively pay attention. Of course, some drivers found the autopilot feature so trustworthy that they let it drive the car while they read or watched a movie, and some drivers got into crashes and were killed because the autopilot system was not yet foolproof.

Musk became notorious for his tendency to tweet on Twitter whatever he was thinking, sometimes leading to damage to his personal reputation. Yet, he continued to innovate and had a loyal following. By 2022, over 2,300 low-orbit Starlink satellites for internet access had been launched by SpaceX. He founded the Boring Company in 2016 to dig tunnels for

Electric Tesla vehicles recharge. CEO Elon Musk described the Tesla Model S as a "very sophisticated computer on wheels." (Mikephotos/Dreamstime.com)

high-speed underground transport. The goal was to build Hyperloop tunnels (straight out of science fiction) hundreds of miles long for fast transportation. The boring machines were automated and computer controlled. Musk and his partners also began building Gigafactories to drive down the cost of manufacturing batteries for Tesla cars and solar power systems. The first Gigafactory opened in 2016 in Nevada and was the second-largest building in the world at 1.9 million square feet.

Musk's vision of the future that he was helping to build was not all optimistic. He had voiced his concern that AI technologies must be regulated to prevent an AI from turning rogue and superseding humanity. He considered AI more serious than nuclear weapons. In 2019, he said that he was concerned by the thought of "humanity as a biological boot loader for digital super intelligence . . . like the minimal bit of code necessary for a computer to start. Like you couldn't evolve silicon circuits—there needed to be biology to get there." In response, he invested in ways to augment human intelligence, such as the 2019 start-up Neuralink, which aimed to build brain implants. Many other technology innovators found Musk's concerns overblown, but Musk had become a celebrity as well as an innovator, lending his concerns more publicity. While electric cars and spaceships technologically pushed humanity forward, a number of critics noted that a great divide was growing between the digital haves and have-nots.

## THE DIGITAL DIVIDE

The *digital divide* (sometimes described as a lack of *digital inclusion*) refers to a division between those who can and cannot access technology, such as the internet, or the technology to utilize the internet. A set of studies beginning in 2000 found that more than half of the computers on the planet were found in the United States, where 51 percent of U.S. families owned a computer and 41 percent of U.S. homes had internet access. Western Europe was also well connected, with 61 percent of Swedish homes having internet access, though only 20 percent of Spanish homes had internet access. Thirty-three percent of Asian homes had internet access, especially concentrated in Japan, South Korea, and other up-and-coming economic powerhouses. Many nations, especially in Africa, had only a minuscule number of computers and limited internet access. These statistics illustrated the global digital divide, where citizens of affluent countries had access to computers and the internet and poorer countries were not able to provide the education and infrastructure to compete in an increasingly globalized economy.

The same set of studies in 2000 found that in the United States, 46 percent of white Americans had internet access at home, while only 23 percent of African American and Hispanic homes had access. Eighty-six percent of households that earned more than $75,000 per year had internet access, and only 12 percent of households that earned less than $15,000 per year had access. Sixty-four percent of households with a college graduate had internet access, and only 11 percent of households had access if no one in those homes had ever graduated from high school. Urban and suburban homes were more likely to have internet access than rural homes or homes in impoverished inner cities. This showed that the digital divide also existed within nations and was exacerbated along racial, income, geographic, and educational divides.

Observers were concerned over access to computers and the internet as well as the knowledge of how to use them. Some argued that computer literacy was almost as important as traditional literacy in order for a person to fully participate as a citizen in a modern democracy or even to find a good job. Various efforts were made to bridge the digital divide. These concentrated on educational efforts and government programs to provide more universal access to the internet—similar to earlier programs to provide universal access to basic telephone service.

Reports continued to show the impact of lack of access through the 2010s. Different access contributed to differing economic, educational, and health outcomes. Numerous stories told the overall tale. Many children did homework riding on Wi-Fi–enabled school buses or sat outside their schools to gain access to the school's Wi-Fi because they did not have access at home. According to one 2016 study, almost half of all students were unable to complete their homework because of lack of access. Another study showed that the parents of these children were looking for employment from inside their cars while parked outside of restaurants after hours, researching opportunities, filling out job applications, and uploading resumes.

In the twenty-first century, the U.S. Bureau of Labor Statistics showed a strong correlation between unemployment and lower broadband access. Basic services like paying bills and setting doctor's appointments assumed internet access. Beginning in 1985, to varying degrees of success, the Federal Communication Commission (FCC) alternately proposed and ran programs to assist Americans below the poverty line to afford both telephone and internet services. Beginning in the 1990s, libraries saw themselves as the stopgap in this division and increased the number of computers available to patrons.

In 2019, the FCC indicated that only 24.7 million of 237 million Americans lacked access at home. However, in 2015, the FCC had declared 25 Mbps as a minimum for broadband, and a 2017 Pew Research Center report argued that many, almost 162.8 million Americans, had internet access but not at broadband speeds. These Americans were mostly in rural areas, but it was also an issue in the downtrodden inner-city areas of cities like Detroit. Only 55 percent of people in rural areas had broadband speed connections compared to 94 percent in urban and suburban areas. Tribal lands had closer to 40 percent with access.

In some locations, wireless communication over cell networks had replaced wired homes because cell speeds had increased faster compared to wired connections. Geography had driven cell connection usage as well. A single gigabyte of data used by mobile devices, or anything transmitted over a cell network, had dramatically different costs based on where you were located. In February 2020, in the United States, transmitting data cost an average of $8 per gigabyte, one of the more expensive in the world. The worst prices were found in Africa and some small island nations; for example, Benin and Malawi cost $27 per gigabyte. The best price in the world was found in India, a mere $0.09 per gigabyte. India had vigorous competition between cell providers. The average price in the world was $5 per gigabyte.

A number of technological approaches to the digital divide issue were pursued. Various efforts by major players were made to reach rural locations by using unused portions of the radio spectrum (Microsoft) or by putting up long-lasting solar-powered drones (Facebook) or balloons (Google) to transmit internet radio signals. In 2018, SpaceX began to launch microsatellites, often piggybacking them on commercial launches for paying customers, in an endeavor to build the Starlink satellite constellation project to deliver cheap internet to the whole planet. Even so, some measurements, such as kilobytes per second per user, indicated that the network divide continued to widen, not shrink, even in 2022. Only three countries, China, Japan, and the United States, used almost 50 percent of the global bandwidth in 2020. Because of poverty and a lack of opportunity, most of the nearly eight billion people on the planet did not have access to computers other than cheap cell phones.

In 2005, Seymour Papert (1928–2016) and Nicholas Negroponte (1943–) at the MIT Media Lab created the nonprofit One Laptop per Child (OLPC) initiative with the intent of creating and distributing $100 laptops to children in third world countries. They designed the XO-1 to have low-power usage.

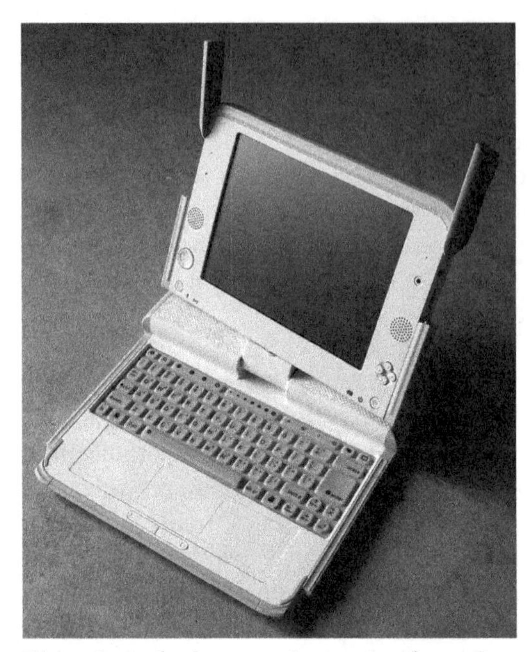

The relatively inexpensive and robust One Laptop per Child (OLPC) initiative XO-4 laptop. Designed to democratize computing worldwide, millions were distributed. (Peanutroaster/Dreamstime.com)

The original design even included a hand-cranked generator to charge the laptop's battery. It used an open-source operating system based on Linux, communicated with an 802.11 mesh network to link laptops, and came in a hardened and water-resistant case that also contained a partial copy of Wikipedia on the hard drive. Uruguay became the first country to order the laptop in 2007. According to OLPC, over three million laptops, including the newest XO-4, were distributed by 2015.

OLPC did not operate without hiccups, however. The computer's cost remained stubbornly higher than $100, and the purchasing countries were responsible for maintenance and figuring out how to best use the computer. Many laptops failed. The organization sold to governments, and critics argued that precious resources allocated to clean water, books (which were less costly), and health efforts like battling ringworm would serve children better. In addition, by 2020, so many older desktops and laptops existed in the world that nonprofits could distribute those at lower costs. A 2013 study in Uruguay showed no improvement in literacy or math and that most of the machines were used for recreation. The laptop did inspire the cheaper Chromebook and the under $50 Aakash tablet computer produced in Hyderabad, India, as comparatively inexpensive educational tools.

While many in the world had no access, many critics also noted that for those that did, the services came with costs for consumers and for those working for the services. In the United States, delivery people for Uber and DoorDash were not considered employees but contractors. Both companies found ways to not offer benefits or share tips with their workers. Meanwhile, the industry's wealthiest companies—specifically the "Big Five" of Amazon, Apple, Alphabet (Google), Microsoft, and Facebook (which changed its company name to Meta in 2021)—absorbed smaller companies and accounted for a huge percentage of the New York Stock Exchange. Although this paralleled general market consolidation in the United States—the top 100 public companies went from 49 percent to 84 percent of earnings between 1975 and 2020—these technology companies in particular came under scrutiny in 2020 for monopolistic practices. It was hard to engage in the modern economy without using, at various levels, their products.

# TEN

# Information Security

## HACKING

The mental image of a hacker as a malicious character, derived from news media, movies, and cyberpunk novels, became affixed in the public imagination in the 1990s. The term *hacker* first emerged in the 1960s, when writing computer code was a difficult esoteric art and computer experts were called hackers as a badge of honor. Some of these hackers used their skills to enter systems without permission. Another breed of technical wizard emerged in the 1960s with *phone phreaking.* Phreakers learned how to manipulate the AT&T phone system to avoid payphone charges, make prank calls, and make free long-distance calls. "Captain Crunch," the most famous of the phreakers (and who we saw earlier as the creator of EasyWriter), earned his name when he discovered that the whistle that came in a box of Cap'n Crunch breakfast cereal emitted the right frequency to subvert the AT&T phone system. At that time, telephones were based on acoustic technology and responded to different tones to know what numbers to dial, whether enough change had been deposited into a payphone, or how to activate the repair mode that telephone repair personnel used to troubleshoot problems. Captain Crunch reputedly taught Steve "Woz" Wozniak how to be a phreaker, and Woz got his friend Steve Jobs involved in their first business: building electronic "blue boxes" to sell to other phreakers to cheat the phone company. Of course, Woz and Jobs later turned honest and founded Apple Computer.

An early example of hacking was discovered in 1986 when a system administrator named Cliff Stoll (1950–) at Lawrence Berkeley National

Laboratory in California was assigned to discover why the system accounting software had a seventy-five cent error. Stoll was an astronomer by profession but working as a UNIX system administrator as a paying job. (This was a common situation. Scientists and engineers picked up extensive computer skills and ended up working in the more lucrative computer field because jobs in their chosen professions were scarce or they just liked having more money.) Stoll found that someone had gained unauthorized access to the laboratory system and was using its system to log into other systems. The time used had not been assigned to a regular user, leading to the accounting error.

Stoll was curious and took advantage of the fact that most intersystem communication at the time used the command line, so all he had to do was hook up a printer to the communications line and print out the commands of the hacker and the results of the sessions. The hacker was quite industrious, often running multiple sessions at a time on separate communications lines, so Stoll borrowed more printers to track his quarry. He wanted to figure out who this person was, so he started to trace back the connections, which he could do because the hacking sessions took hours and the hacker returned regularly. Because the hacker had daisy-chained himself through numerous systems, Stoll had to contact the owner of each system in turn and then recruit the administrator to help him in his quest, tracing back the next link. The hacker also surreptitiously used Mitre, Tymnet, and Datex services, which provided banks of modems for long-distance and overseas telephone connections.

The hacker was obviously interested in military information, especially information on the Strategic Defensive Initiative (SDI). This effort was nicknamed "Star Wars," a multibillion-dollar effort by the Reagan administration in the mid-1980s to build antimissile defenses. The hacker used Stoll's system to log into other systems on the Defense Department's MILNET, such as the Anniston Army Depot in Alabama, continuing his search for national security information. Stoll contacted law enforcement. The Federal Bureau of Investigation (FBI) was befuddled as to why Stoll was so obsessed with a seventy-five-cent error. Computer hacking was not yet an issue for law enforcement. Stoll continued on his quest, driven by an obsessive itch to solve the problem, and like a well-trained scientist, he kept an extensive logbook.

Stoll contacted the Central Intelligence Agency (CIA) and the National Security Agency (NSA), and while both organizations were interested and wanted to be kept up to date, they were not in a legal position to handle this type of situation. He kept the Department of Energy informed as well as the Air Force Office of Special Investigations, as the national laboratory that Stoll worked at was funded by the Department of Energy and also designed nuclear weapons.

To keep the hacker online longer, Stoll recruited his friends, and they created a fake department at the laboratory and hundreds of files full of fake information about SDI and other military secrets. The hacker eagerly downloaded these files, which took a long time on 1200-baud modems. Stoll had stumbled across a technique that was later named a *honeypot* or *honeynet* by information security professionals. Stoll eventually tracked down the hackers to a small group of young men in Hanover, West Germany, who were freelancing and selling what they found to the Soviet KGB. Though amateurs, they were spies, who were arrested and convicted; the men had not succeeded in finding any classified information and were given light sentences.

Stoll wrote a popular book about his experiences, *The Cuckoo's Egg: Tracking a Spy through the Maze of Computer Espionage* (1989), which eventually sold over a million copies. While Stoll did not continue to work in the emerging information security field, following other esoteric interests instead, his book inspired numerous other people to go into the field.

As time went by and some hackers started to cause problems, the term *hacker* changed from a moniker of respect to a label defining an antisocial person bent on making trouble. Some hackers, wanting to regain the title, suggested that the term *cracker* be used for the bad guys instead, but that never caught on. Some alternate terms that did catch on were *white hat* hacker and *black hat* hacker, recalling the days of cowboy movie serials where the audience never doubted who was good and who was bad.

Many hackers were motivated by the desire to seek knowledge and to be respected for their technical prowess, even if their real identity was hidden behind a pseudonymous handle or they were part of a larger hacker organization. An example of one such group was the semipublic Cult of the Dead Cow in the 1980s and 1990s, who became known for *hacktivism* (hackers as social or political activists) and released free hacking software, such as Back Orifice in 1998. This Trojan horse software (malicious software that hid inside innocent-looking software) allowed a user to remotely control a Windows 98 machine.

Serious hackers enjoyed conquering difficult technical problems and were often thrill seekers or were seeking revenge. Employees who had been laid off were a major source of hacking and continued to be an important information security threat. The best hackers often sought financial gain, either through industrial espionage or outright theft. Multiple episodes were reported where hackers stole credit card information from a bank or online retailer and blackmailed the targeted company for the return of the information. Companies often paid the blackmailer because they wanted to avoid a public relations disaster and loss of confidence by their customers. Organizations dealt with far more hacking than the public knew about. Hacktivists in the 1990s occasionally defaced websites of corporations or organizations that they disagreed with, and in the 2000s and 2010s, hacktivists became more effective by obtaining and releasing embarrassing internal information.

So how did hackers do it? Quite frankly, social engineering often proved to be the most effective method. For example, a hacker might call up someone in a company posing as a technician from the internal computer support department and ask for a password. Much of the time, the hacker got the password through this simple ruse. One of the most famous social engineering hacks occurred in 1978 when a computer consultant named Stanley Mark Rifkin performed a temporary assignment at the Security Pacific National Bank in Los Angeles. He learned how the system of wire transfers worked and realized that security centered around a daily code that was given over the phone by bank executives to authorize a wire transfer. He visited the wire transfer room, saw the code written on a piece of paper, and then went outside the bank to a payphone. When he called the wire transfer room, he pretended to be a bank executive and asked for $10.2 million dollars to be transferred to a bank account in Switzerland. When asked for the code, he gave them the number. Rifkin then flew to Switzerland, converted the cash in his account to diamonds, and returned to the United States to enter *The Guinness Book of World Records* with the temporary record for the "biggest computer fraud."

Hackers could also use a program called a port scanner to knock on all the doors in a network or computer system to see whether any points of entry were willing to communicate. After gaining communication with a system, the hacker still needed to use his or her bag of tricks to try to gain access to the system. Besides their own custom software, hackers could often search the many hacking databases available on the internet for an exploit that opened the vulnerable hole wide enough to gain control of the target system. The best hackers came and went with no one ever being the wiser. The internet, by connecting computers around the world and creating a massive cybernetic organism, made it easy for hackers to hit computer systems or networks from another continent.

The most effective attacks did not come from the brute force of crashing systems but subtle changes in data on computer systems that led people to make bad decisions. Possible attacks on digital control systems for gas pipelines, electrical power grids, hydroelectric dam controls, sewage treatment plants, water distribution systems, oil refineries, or chemical manufacturing facilities threatened to cause major utility disruptions and environmental damage.

The key to whether a computer system was vulnerable was whether the hacker could communicate with the system. A hacker had to interfere with a wireless signal (found in links with satellites or by using microwave, infrared, or radio waves) or have access to the actual communications wire that was attached to a computer system. The U.S. military and different government agencies separated computer networks from the open internet, though even these private networks were often created by leasing network bandwidth from large national and international telecommunications providers. The only way for a computer system to be completely safe from a hacker was to keep the computer in a locked room and not have a network.

The terms *computer virus* and *computer worm* refer to different things. Just as in biology, a virus needs a host to live in and reproduce, where a worm can reproduce and travel about on its own. Perhaps the most famous of all worms, the Morris worm, happened in 1988. Robert Morris Jr. (1965–), the graduate student at Cornell University who wrote the worm, did not intend for it to damage anything; he just wanted to see how far his worm would spread. A bug in the program allowed the worm to keep reinfecting systems, and it eventually caused systems to slow down and sometimes crash. The internet was estimated to have approximately 500,000 users at that time, and the Morris worm infected an estimated 10 percent of all machines on the internet. The federal government convicted Morris of his crime and sentenced him to three years of probation and a fine. As the twentieth century ended and the new century began, the public became increasingly aware of the impact of computer viruses and worms, with exotic names such as Slammer, Nimba, Code Red, and Melissa. Most of these viruses and worms in the 1990s and early 2000s were relatively unsophisticated efforts, though they caused the loss of millions of dollars in lost computer time and needed cleaning up afterward.

For information security professionals, the white hats, the best defense came from knowing your enemy. Some people proposed counterattacking any attacking machine, getting the hacker before they get you. However, it was discovered that any hacker worth their salt launched their attacks from computer systems that they had already compromised, and the true owner of the system was unaware that their system was being used illicitly. Security professionals, sometimes called *ethical hackers*, always got permission from the proper authorities before doing a security analysis of a computer system or network. This permission was referred to as a "get out of jail free card," because would-be security analysts went to prison while claiming in their own defense that they just wanted to demonstrate security problems that existed in an organization.

Good security meant an absence of problems through diligence, an intangible that was hard to justify spending extra time or money on in the 1980s and 1990s. Most people tended to think about security as an afterthought, though that changed after decades of publicity about computer hackers. The attitude of the FBI, for example, changed after the 1980s and the episode with Cliff Stoll; a Cyber Division was created in 2002 with extensive resources, and the FBI began publishing a "Cyber's Most Wanted List." Most of the people on the list in 2020 were foreign nationals, often from Nigeria, China, Eastern Europe, or Russia.

## PUBLIC-KEY ENCRYPTION

Before the 1970s, all encryption/decryption algorithms were symmetric, in that they used the same algorithm and digital key to both encode and decode a message. In 1975, a pair of American cryptographers, Whitfield Diffie (1944–) and Martin Hellman (1945–), developed the mathematics for public-key encryption, an asymmetric algorithm. Two years later, Ron Rivest (1947–), an American; Adi Shamir (1952–), an Israeli; and Leonard Adleman (1945–), another American, developed and patented the RSA algorithm (named after their initials), an effective implementation of public-key encryption that became a standard in the field. In its simplest form, public-key encryption allowed for the creation of a pair of digital keys that could then be used to encrypt or decrypt a message. Only the opposite key can decrypt what the other key of the key pair has encrypted. Possessing only one key did not allow someone to break the encryption.

Public-key encryption also led to another innovation: digital (or electronic) signatures. Developed in the late 1980s, the process of creating a digital signature started with a person creating a connected pair of keys. One key was called the *public key* and was given to the world. The other key was called the *private key* and was kept secret. To create a digital signature, a person ran a program that analyzed the document to be signed and created a hashed number that uniquely described that document. (A hash was a recipe to convert a large amount of data down into a smaller number that was unique to that set of data; if the data was changed by even a single bit, the recipe would return a different number when it was run again.) The private key was then used to encrypt the hashed number, creating a series of bits that were the unique digital signature for that document. The signer then sent out the document with the digital signature and was legally bound by that signature, just like a physical signature on a legal document. In order to verify that the digital signature was accurate, another person just had to decrypt the signature using the public key of the signer and then run a program to get a hashed number for the document. If the new hashed number was the same number that was held in the digital signature, you knew that the document had not been altered and that it could only have been signed by the person who possessed the private key of the signer. It was a clever technology that sounded complex, but the complexity was hidden inside programs to manage digital signatures. Utah was the first state to pass a law making digital signatures legal in 1995, an effort to encourage the technology to grow. The European Union made such signatures legal in 1999, the United States followed suit in 2000, and the United Nations issued model legislation for member nations in 2001.

## HACKING EVOLVES

By the 2010s, the two most serious cybersecurity problems in the world were cyberattacks by well-funded nation-states and cyberattacks by organized criminal groups. The day of the individual hacker was being eclipsed, replaced by teams who built complex malware (malicious software) packages. New terminology become more prominent to describe specific hacking techniques or information security problems. For instance, a *zero-day* attack was an undiscovered flaw in an operating system or program that had not been revealed before. These techniques were the magical keys of the hacking world and closely guarded. Illegal cybermarkets for selling such attacks formed, and nation-states often purchased such attacks and kept them in their war chest for future attacks. Access to these black cybermarkets often occurred through the dark web.

The dark web was a useful concept but not a single organized thing. In essence, the dark web consisted of websites that were not visible to search engines and chat rooms and other communication channels that were not readily available to normal users. The concept of the deep web also existed, websites and data that were not visible to search engines, often because such sites were behind paywalls or the data was inside databases that were not readily searchable by the web spiders that search engines used to compile their indexes. The dark web, contrarily, was used to facilitate illegal activity.

Sophisticated users of the dark web communicated using software that provided encryption and networks (like the Tor network) that were encrypted. Their goal was to not be detected by governments. Governments and network companies could detect that encrypted traffic was flowing but could not readily read the contents of the traffic. One of the conclusions apparent from the NSA tools that Edward Snowden (described in more detail below) revealed in 2013 was how effective encryption was; most of the tools were designed to grab data while it was still cleartext (human readable text), before it was encrypted or after it was decrypted. All data at some point needed to be in cleartext; otherwise, it would never be useful to human beings. The exception to that rule was data in the form of encryption keys or digital signatures that remained encrypted because it was not human readable but only read by encryption software to verify that accuracy of human readable text.

Illegal cybermarkets were often found on the dark web, which used cryptocurrencies to sell illegal drugs, malware, illegal pornography, and other such material. Silk Road was one such market, selling illegal material from many vendors before being shut down in 2013 by a federal investigation after two years of operation. The programmer who founded Silk Road called himself "Dread Pirate Roberts" (a reference to the novel and movie *The Princess Bride*) and was sentenced to life in prison. The site was sophisticated enough to provide an escrow service, where bitcoins were held by Silk Road for a buyer until the illegal material had been delivered or an illegal service performed, a clever way of creating a degree of trust between untrustworthy people. The federal government later sold the seized bitcoins for millions of dollars.

Other hacking approaches included phishing and spear phishing, both made less dangerous by dual-factor authentication. *Phishing* was a variation on social engineering, noted earlier, but the approach was sending an email or text claiming to be a trusted source and asking the recipient to do something. A common example of this was telling the victim that they needed to log into their banking site and verify their account information.

A false link to their banking website was helpfully provided, and clicking on the link led to a fake website that looked like the real website. The victim then logged in, giving their username and password to the fake website. The criminal then used those credentials to log into the real bank website and transfer funds to another account that they could then withdraw the money from. Phishing attacks were usually sent on a massive scale, even millions of emails, using email spam lists. Only a small fraction of people had to fall for such scams to make them extremely profitable. An early study of phishing found that one in twenty people fell for the fake information.

The most effective solution to phishing was dual-factor authentication, a specific type of multifactor authentication. *Authentication* was the process of persuading a computer who you were. This was done by using something you knew, something you had, or something you were. Examples of something you knew were usernames and passwords. Examples of something you had were a physical key, a card with sequential numbers on it that are used one at a time with each login, or a number sent to a smartphone application that was valid for only a short period of time. Examples of something you were included fingerprints, eye scans, handprints, or other biometric measures. Using only one of these three categories of things was called single-factor authentication. Even using both a username and a password was still single-factor authentication because they were both things that you knew. An example of dual-factor authentication was using a password and your fingerprint, which were from two of the three categories. Serious sites, usually rooms with restricted access, used triple-factor authentication.

Dual-factor authentication was expensive to implement until smartphones became ubiquitous, and by 2020, many people had smartphone apps that sent them a temporary authentication number to use along with a username and password when they logged into their banking site. Because a fake banking site, used so often in phishing scams, could not complete the login without the temporary authentication number, dual-factor authentication solved much of the phishing problems. Federal regulatory agencies compelled banks and other financial institutions to use multifactor identification, though the banks and financial institutions resisted full dual-factor authentication until smartphones apps made it easy. Instead, they used multiple single-factor authentication techniques, which was using another example of something you know to go with your account number and password. The method they commonly used was displaying a picture during the login that the account holder had previously chosen. A fake website had no way of knowing what the chosen picture was. If the user noticed that the picture was wrong, they would know that they were being deceived. While this system worked in theory, banks and financial institutions did a poor job of explaining the technology, perhaps because they did not want to alarm their customers that malefactors might be trying to steal their banking information.

*Spear phishing* was like phishing, but instead of relying on a large number of targets, spear phishing targeted a small group of people, often a single person. The person writing the spear phishing email or text carefully researched the target, making sure that they were using correct terminology and trying to make the message seem to come from within the victim's circle of trusted sources. Spear phishing was used by Chinese hackers to crack into Google in 2009. Their goal seemed to be accessing the Gmail accounts of Chinese human rights activists, though it later became apparent that the hackers had gained access to a database of court orders that Google was obeying. These were search or surveillance warrants for individual Google users. The hackers also apparently copied the source code used to run Google; examples of what could be found in that source code were proprietary algorithms that allowed Google to efficiently distribute data around the world and how its search engine worked. That Google engineers—some of the sharpest and most savvy engineers in the world—could be fooled by spear phishing emails was a sobering reminder of how easy it had become to steal intellectual data and subvert an individual's circle of trust.

A *bot* was a program that acted as a digital robot, usually running scripts that performed a digital task over and over again. A bot might be a program used in a game that pretended to be a user doing something repetitively, liking mining for gold in *World of Warcraft*. Malicious bots were used to launch denial-of-service (DoS) attacks, where a program tried to overwhelm a computer across a network in some way, such as repeatedly sending malformed network packets that consumed a lot of CPU cycles to process. A distributed denial-of-service (DDoS) attack used bots on many different systems to attack the same target simultaneously. A group of bots coordinated to act in concert was called a *botnet*. There were documented botnets with millions of compromised machines in them. Such botnets were often used to send spam emails or launch very large cyberattacks.

In the early days of botnets, a fifteen-year-old Canadian hacker named "MafiaBoy" used a DDoS attack to knock major sites like CNN and eBay off the Web in 2000. In 2007, Russia launched a massive DDoS attack on Estonia, a former republic of the Soviet Union, which had computerized much of its government functions and was especially vulnerable. Other such Russian attacks were later launched on Ukraine, and a similar attack was launched on Georgia in 2019. These countries were all former Soviet republics that had tried to chart international paths away from Russian influence.

A 2016 DDoS attack managed to knock Netflix, Amazon, and other major sites off-line by attacking a DNS (domain name system) provider for those companies. The motive remained unclear, though the attack was notable because the botnet used in the attack had many computers in it from the Internet of Things (IoT) instead of the normal majority of home personal computers (PCs). Users often treated IoT devices as a tool that they could set up and forget and did not need to maintain. The manufacturers of IoT devices also often ignored the fact that they had created computers that could be compromised to run malware. Security updates to the IoT systems were hardly ever provided nor even simple ways to update the software.

## HACKING AS CYBERWAR

Discussions of information security often turned to the nightmare scenarios of cyberwar, which could often be both scary and confusing. Beginning in the 1990s, national security experts and pundits regularly warned of a cyber Pearl Harbor, an event like Japan's infamous 1941 attack on the U.S. Navy, where the United States would be caught unawares but this time suffer dramatic damage to power networks, computers systems, robots, and every other technology that heavily depended on computers.

In 1997, the National Security Agency (NSA) authorized a team of its hackers to penetrate American military and civilian computers in an exercise called "Eligible Receiver." It took several days before the unsuspecting military believed that they were really under attack. While the particulars have not been released, apparently the hackers experienced considerable success, leaving messages on systems they had compromised, taking over command center computers, systems on power grids, and the 911 emergency call systems in nine American cities. Even over two decades later, the documents at the National Security Archive on this exercise released through Freedom of Information requests had much of the contents redacted.

In 2000, as part of the security effort for the 2002 Winter Olympics in Salt Lake City, the U.S. Department of Energy and Utah Olympic Public Safety Command ran an exercise called "Black Ice." Hundreds of participating officials tried to cope with a faux event: a major ice storm that had damaged power lines and resulted in a policy of rolling blackouts in response. Officials were surprised how much intermittent power degraded their infrastructure and affected their ability to communicate in an emergency. While the scenario did not deal with cyberterrorism, the effects demonstrated how interconnected the various systems of the public infrastructure were and how vulnerable they were to malicious attack. A real example occurred that same year when, after at least forty-six attempts to break into a computerized sewage system in Maroochy Shire, Australia, a disgruntled former employee succeeded and released one million liters of raw sewage into local parks and streams.

Cyber aggression between nation-states ranged on a continuum from simple spying (cyberespionage) to cyber influence, then cyber sabotage, and, finally, to full cyberwar. Spying, what nations had been doing for thousands of years, had extended into the realm of computers and networks in further attempts to gain useful information. An example of this was the Five Eyes, an intelligence alliance formed after World War II between the United States, United Kingdom, Canada, Australia, and New Zealand. These nations cooperated to spy on the wired and wireless communications of the rest of the world, an important part of the Cold War against the Soviet Union and world communism. The Echelon system became a key part of this effort and was publicly disclosed to the European Parliament in a 2001 report. It revealed that the Five Eyes had listening posts all over the world to monitor communications via cable, radio, satellite, or internet.

*Cyber influence* referred to using social media and other forms of media to influence public opinion or to shape the intellectual environment and idea marketplace. It could be seen as an extension of government-sponsored propaganda. The content being spread may have been true or false, although false narratives, called *misinformation*, was more commonly found in such cyber influence campaigns. An example described later in this chapter was the Russian attempt to influence the 2016 American presidential election. Other terms for cyber influence campaigns were *cyber meddling* or, an older term, *PsyOps*, which stands for *psychological operations*, a form of propaganda usually created by military units. Cyber influence attacks included efforts to spread disinformation and exert foreign influence through YouTube video channels and Google advertising accounts. Facebook, which we will return to shortly, may have been the most perfectly designed platform ever for use by PsyOps. Google created a Project Zero in 2004, which became the Threat Analysis Group (TAG), to counter government-backed cyberattacks; many other companies created similar teams.

Beyond espionage or influence, *cyber sabotage* actively tried to physically damage another nation or to obtain sensitive information that was then publicly released. Sabotage was a step beyond just cyber influence. An example was when North Koreans hacked into Sony Pictures in 2014. The company planned to release a comedic film, *The Interview*, that mocked the dictator of North Korea, Kim Jong-un. The hackers erased data and generally caused havoc in Sony Pictures and also stole the content of emails, corporate information such as contract details and salary numbers, digital copies of unreleased films, and other confidential corporate data. Though Sony Pictures scaled back the release of the film, restricting screenings to only a few theaters and sending the film directly into digital release on streaming platforms and video discs, the hackers released the corporate information, causing Sony Pictures considerable embarrassment and reputational damage—for example, the revelation that female stars were paid less than male stars for comparable roles. The Sony Pictures hack had an additional advantage for North Korea in that it might dissuade corporations from crossing Kim Jung-un again out of fear that similar hacks might release their own corporate secrets. Another example was the Russian hacking of the Democratic National Committee in 2016 with the intent of releasing the stolen data to change the outcome of the election. That hack is described below.

*Cyberwar* (also called *information warfare* or *netwar*) used computer hacking as another tool in actual warfare for both defense and offense. Cyberwar grew out of two twentieth-century developments in warfare. As troops grew to rely on radio, radar, sonar, and other electronic sensors, enemy forces developed electronic countermeasures (ECM) to confuse those electronic sensors and even render them useless. This could be as simple as jamming a radio frequency with noise to deny that frequency to adversaries for radio transmissions. For that reason, more modern soldiers began using frequency-hopping radios that moved so quickly from frequency to frequency (a technique patented by composer George Antheil (1900–1959) and actress Hedy Lamarr (1914–2000) in 1942) that the enemy could not keep up and adjust their jamming equipment quickly enough. The effort to counter ECM had its own acronym, electronic counter-countermeasures (ECCM), and the war within the electromagnetic spectrum created a never-ending cycle of creating new attacks and new defenses.

The second historical development included ever more sophisticated communications, command, control, and intelligence (C3I) military infrastructures. Modern warfare, as practiced by the American military, relied extensively on computers and electronics. Satellites spied on the enemy and transmitted encrypted communications, computerized databases streamlined logistics, and the American military began deploying electronics so that each soldier was literally turned into a node on a network. This empowered soldiers to be able to feed video back to their commanders, receive orders, and view video feeds from small overhead drones. The ultimate cyberwar weapon was probably a nuclear device modified to maximize its electromagnetic pulse (EMP), which could literally melt running electronics and electrical systems with its energetic particles.

The requirement for the American military to always be able to precisely locate its personnel and equipment anywhere on the earth's surface led to the Global Positioning System (GPS). This system, originally a system of twenty-four satellites launched between 1989 and 1994, allowed anyone with a GPS receiver to locate themselves on the surface of the earth with closer than ten meters accuracy. GPS equipment attached to missiles and bombs made precision weapons easy and effective, seen most prominently in the wars in Afghanistan, Iraq, and Syria. Ever since the 1920s, aerial bombardment enthusiasts expected air power to become the decisive weapon on the battlefield, and GPS-enabled precision weapons may have actually fulfilled that promise. Anyone could use GPS signals, and GPS receivers revolutionized the practice of scientific fieldwork by allowing precise measurements of continental drift and the locations of geological formations, archaeological sites, and animal populations. GPS receivers were used by hikers, drivers, and sailors and used to locate stolen cars and track commercial shipments. Many smartphones contained GPS receivers, providing the accurate location revolution to common consumers. Potential enemies could also use GPS receivers, and the American military was rumored to have a feature that, if necessary, would change the GPS signals so that only American military GPS receivers would continue to work and all civilian receivers would fail.

While people usually use the term *GPS*, a more formal term is *global navigation satellite system* (GNSS), because other nations, not wanting to be dependent on the American system, built similar systems. China's system was initially called Compass but became the BeiDou Navigation Satellite System (BDS). The European Union created a version called Galileo. GLONASS was launched by and controlled by Russia. Smaller localized systems were created by both India and Japan.

Many American federal agencies had overlapping responsibilities on the issue of cyber national aggression, including the Pentagon, the NSA, the FBI, and the CIA. The NSA, nicknamed the "puzzle palace," was perhaps the most misunderstood of the agencies. The NSA, founded in 1952 by presidential order (the only federal agency founded without congressional action), was designated the codemaker and codebreaker for the U.S. federal government. The goal of the NSA was to conduct electronic surveillance around the world, break the codes if encrypted messages were found, and create encryption schemes for use by the American military and government that could not be broken. The NSA was also rumored to have more supercomputers than any other organization in the world—to better make and break codes.

One of the more successful efforts by the CIA and NSA to listen to the secret messages of other governments came from compromising the encryption machines sold by Crypto AG. During World War II, an inventor named Boris Hagelin (1892–1983) fled Norway and founded a company named Crypto, which manufactured encryption machines for the American military. This M-209 machine was used by troops on the ground and was not as sophisticated as the Enigma machine or other high-level encryption machines. Hagelin moved his factory to Sweden after the war and improved on his product, then he moved his operations to Switzerland. The United States made an agreement with Hagelin that it would pay him to sell weakened versions of the machines so that the NSA could more easily break the encrypted traffic. Crypto AG machines were sold to third world countries, not to the United States or its close allies, nor to the Soviet Union or its close allies, as the two superpowers did not trust encryption products that they did not completely control.

When Crypto AG rolled out its first all-electronic model in 1967, the innards had actually been designed by the NSA. There was no backdoor, just a subtle weakening of the encryption algorithm to help the NSA quickly break the encrypted messages with its powerful computer systems. The CIA and West German intelligence purchased the company in 1970 and remained its secret owner. The two spy agencies even made good profits off their investment. At times in the 1980s, 40 percent of all the diplomatic cables and other transmissions that the NSA intercepted were using Crypto AG encryption machines. In 2018, the company closed.

The major concerns of cyberwar theorists extended beyond just the potential effects on the battlefield or a potential Pearl Harbor cyberattack. More personal scenarios were also easy to imagine: what if you went to your ATM and could not withdraw any money because a bank computer system thought that your balance was zero; or you could not fly because the air traffic control system had crashed; or you could not leave your sorrows behind in a game of *Fortnite* because the internet was clogged with rampaging worms, viruses, and DoS attacks; or you had no electricity because the computers systems controlling the electrical grid had shut down? Various forms of malware could be used to achieve many of these objectives through hacking by criminals or nation-states. For criminals to fully realize the opportunities of blackmail in the digital age, they needed another technical innovation: cryptocurrencies, a way to transfer money semianonymously through computer networks.

## CRYPTOCURRENCIES

*Money* has been taken to mean wealth, a measure of assets, the value of one thing compared to another thing, or as a measure of a means of exchange used to engage in economic transactions. Money has often been a physical thing: bills, coins, or even the conch shells of the thriving Indian Ocean economy of the Middle Ages. However, the twenty-first century saw most money stored as bits in bank accounts in computers. Throughout history, different valuable things had been used to guarantee the value of currency and coins, such as vaults filled with gold, silver, or iron or expanses of fertile land. The gold standard became common in the Western world, based on the idea that if a government issued paper currency, anyone could take that paper currency and redeem it for actual gold. After the United States effectively abandoned the gold standard in 1971, American currency became a fiat currency. That meant that American dollars were recognized as currency because they were backed by the prestige of the U.S. federal government but had no intrinsic value. A vocal minority of economic libertarians had long agitated for a form of fiat currency that was not controlled by an individual government. They thought their dream had come true in 2009.

An unknown person named Satoshi Nakamoto, which was assumed to be a pseudonym, created a sophisticated cryptocurrency that he called *bitcoin* and released it onto the Web in that year. Bitcoin used public and private key technology to create unique digital bundles. Actual bitcoins had to be created through a complex series of mathematical operations, meaning that a person had to expend an enormous number of CPU cycles to make a single bitcoin. Bitcoin mining operations sprang up around the world, especially where cheaper electricity was available to run the server farms dedicated to creating bitcoins. Bitcoins were often treated as an investment by buyers, although they did not reflect a material investment because the digital bits of the bitcoin itself had no value, unlike gold, land, or iron, nor could the bitcoins earn value through charging for a service, producing a physical good, or creating some other useful economic activity. It was essentially investing in a currency that was disconnected from any nation-state. The total number of bitcoins was limited by the algorithm to twenty-one million bitcoins, and many financial transactions using bitcoins were done in fractions of a bitcoin. This created a form of artificial scarcity.

Nakamoto invented blockchain technology to create bitcoin, and in many ways, blockchain was a more intriguing innovation. This was a distributed ledger technology based on using digital signature technologies that recorded transactions. It could be used to create smart contracts, where parties were bound by their contracts because of the use of digital signatures that only that person or company could apply. Other innovators created their own forms of cryptocurrency based on the original blockchain technology ideas and their own algorithms. Such examples included Ethereum, Ripple, Litecoin, and Stellar. These currencies were traded on markets, which often fluctuated widely based on movements of small amounts of coins.

One of the dreams associated with bitcoin and the other cryptocurrencies was to have truly anonymous digital currencies. Anonymity was desired because of libertarian ideals about minimizing government influence in economics. Anonymity also proved to be a boon for another type of economic activity: criminals collecting ransoms.

## RANSOMWARE

In March 2018, Atlanta was struck by hackers in a ransomware attack. A city like Atlanta, Georgia, with a population of half a million people, had thousands of employees and thousands of computers. The police department alone had 2,000 employees. The premise behind ransomware was for a criminal hacker to break into a system or systems and hold the target for ransom. Because a hacker could be evicted relatively easily if a system administrator knew that the hacker was there, the criminal used encryption software to encrypt data on the hacked system or systems. The attack on Atlanta used the SamSam ransomware package, which infiltrated Java-based server environments via the

Remote Desktop Protocol (RDP) that Microsoft had developed for administrators to remotely control other systems. Citizens could not pay their city water bills, police records systems did not work, and courts could not schedule hearings. Atlanta was fortunate in that most of their systems were not affected, although over 3,000 systems were compromised. A detailed ransom note demanded money in return for the decryption key. The ransom for Atlanta was a mere $50,000, to be paid in bitcoins. Cryptocurrencies and their promise of anonymity solved the long-standing problem for those demanding ransoms of how ransom money could be received without the recipient being tracked. In this case, the ransom was kept low enough to encourage the victim to pay.

The advice from the FBI had been that ransoms should not be paid because it just encouraged the criminals to find new victims. That was nice advice in the abstract, but when the victims of the ransomware attack faced the reality that they needed to get their systems back online and get access to their data again, no doubt Atlanta was tempted to choose to pay. It is not public known whether Atlanta paid the ransom. Even if they had paid the ransom, there would have been no guarantee that the decryption key would have actually worked. Even with a ransom paid, a city, company, or organization would still have a lot of extra expense and effort to decrypt all those computers and get the systems up and running again. In the end, Atlanta spent at least $2.6 million recovering from this attack. The city hired expensive consultants to analyze the attack, recovered data from backup or rebuilt data from original sources, and built a new infrastructure that would resist a similar attack.

A federal grand jury indictment later that year accused two Iranian citizens of being behind the ransomware attack on Atlanta. These two men had also attacked Newark, New Jersey; the Colorado Department of Transportation; the Port of San Diego; the University of Calgary; hospitals; and a number of companies. In all, there were over 200 victims in three years, and over $6 million in ransom was collected via bitcoin. Logs of chat sessions that the federal government obtained showed that the two men had also actually created the SamSam malware. Many cybercriminals, not sophisticated enough to write such malware, just used copies of malware that they obtained through purchases on the dark web. Experts called these cybercriminals "script-kiddies," novice hackers who only knew how to run hacking tools and did not understand the underlying technology. But they could still do damage.

The idea of ransomware had been around since at least the 1980s, but ransomware had become a serious problem by 2012 as sophisticated ransomware packages became more available via the dark web for criminals to use. The FBI reported that in 2015, there were about 1,000 such attacks per day just in the United States, increasing to 4,000 attacks per day only a year later in 2016. Many victims were individuals who accidentally downloaded malware or were hacked in other ways and often just paid the ransom. Early ransomware attacks tended to target home computers; the FBI provided examples of ransom notes:

> Your computer was used to visit websites with illegal content. To unlock your computer, you must pay a $100 fine.
> You only have 96 hours to submit the payment. If you do not send money within provided time, all your files will be permanently encrypted and no one will be able to recover them.

As the malware packages for attacking servers became more sophisticated, going from encrypting data on a single computer to complete systems across numerous computers, bigger targets were sought. Hundreds of companies, cities, counties, hospitals, and other organizations fell victim. Part of the reason that ransomware became such a common global problem was that bitcoin and other cryptocurrencies made it easy to internationally transfer funds as a payment without being tracked. A new variation that began to appear was making a copy of all the data before it was encrypted and threatening to release the data to the public. Some of the data may have included proprietary secrets that companies did not want to get out; other data, like in the Sony hack, were embarrassing secrets. Email conversations often contained such damaging information.

The information technology industry had a process called *business recovery*. This was the practice of having backup systems and backing up all data; in an emergency, backup systems could be brought online and the data quickly restored, and users could continue to operate. Such plans were useful for a potential earthquake, fire, equipment failure, or any event that could take down systems. Business recovery was expensive and tedious, and it was easy for organizations to skimp on devoting money to something that might never happen and to neglect regular practice drills to make sure that backup systems actually worked as expected. Being struck by ransomware was just a new threat, and when it happened, an organization found out whether it was really prepared. Many found that they were not prepared.

Ransomware attacks could be irritating in other ways as well, depending on your taste in music. In October 2019, attackers used the ransomware

FTCode to attack targets in Italy. The attack came through malicious emails sent to Office 365 customers. While the malware was cheerfully encrypting the customers' data, the malware also downloaded songs from Archive.org to serenade users with dark metal tunes from the German punk rock group Rammstein.

## HACKING METHODOLOGY

Beginning in the 1990s, information security became increasingly important. Hackers, viruses, worms, and Trojan horses became a major concern for computer users everywhere. Prior to that time, when a customer asked whether a product was secure from hackers, the software company often said something like, "Of course, it's safe. Our encryption algorithms are secure because we don't show them to anybody. We keep everything secret." Security professionals began to recognize that "security through obscurity" did not work. The best encryption algorithms in the world were public knowledge, often described in detail on Wikipedia, where anyone who wanted to could analyze them for weaknesses. It was this process of review that demonstrated the actual strength of the algorithms.

In the 1970s, movies started to use hackers as characters, portraying them as modern-day wizards possessing powerful and secret skills. Some of these movies and television shows even made an effort to be technically plausible, such as *WarGames* (1983), *Sneakers* (1992), *Antitrust* (2001), *The Social Network* (2010), and the *Mr. Robot* television series (2015–2019). Movies and television series that were truly awful in portraying hacking include *Hackers* (1995), *Swordfish* (2001), the *24* television series (2001–2010, 2014), and most movies that portray hacking as a way to advance the plot.

The first two decades of the twenty-first century saw the whole field of information security change dramatically. In some ways, the previous two decades before the turn of the century had been a time of naivete and children playing with toys. That was a time of phreakers and solitary hackers, who mostly were not acting with malice, but only wanted to show that they were smart enough to figure out arcane technologies and use them to their advantage. When computers were being used literally everywhere and networks allowed those computers to be accessed from anywhere else in the world, the older centers of accumulated political and economic power started to pay serious attention. National governments, military organizations, corporations, and organized crime syndicates began to use hacking to advance their own agendas. It also meant that these organizations were vulnerable to being hacked in turn. Intelligence and counterintelligence (better known as spying and catching the spies, respectively) had a whole new field to master and manipulate. These activities already existed before the year 2000, but the scale changed as ubiquitous computers and networks became an intimate part of what it meant to be human.

Government efforts to thwart hacking and increase information security grew. The federal government also created organizations to help companies and individuals with information security or added those responsibilities to existing organizations. The National Institute of Standards and Technology (NIST), which had been around since 1901, helped American industries both cooperate and compete. NIST guidelines were useful for information security professionals by promoting best practices and better technologies. The United States Computer Emergency Readiness Team (US-CERT) was founded in 1988 by the federally funded Software Engineering Institute at Carnegie Mellon University in Pittsburgh. This organization coordinated responses to computer attacks and malware and provided a catalog of known computer security vulnerabilities in all their variations. US-CERT became part of the Department of Homeland Security (DHS) in 2003.

As the internet grew in importance during the twenty-first century, with more and more people using it for work, play, and interpersonal connections, there was a regular drumbeat of data losses by large private companies. While many companies wished to avoid revealing such embarrassments, the losses were often too large to conceal, as customers started to complain that their personal information had been found on the internet or used in identity theft. California state law required companies to publicly reveal such losses, and almost every company had operations in California (one-seventh of the American economy) and so were required to obey California law, even if the company did not have its headquarters there. Among the victims were Adobe in October 2013, losing 153 million user records; eBay, which lost control of 145 million user records in May 2014; and Marriot International, which lost 500 million customer records from 2014 to 2018. Adult Friend Finder, a site providing pornography and hookup opportunities, discovered in 2016 that it had lost control of information on over 400 million accounts for the prior twenty years. But Yahoo set the record for the largest data breach yet known, information on 3 billion user accounts, in 2013 and 2014, though the company failed to realize the extent of the loss for several years.

An Indian government database, Aadhaar, was compromised in 2017 and apparently lost copies of personal information on almost every Indian citizen, over 1.1 billion individuals, including copies of their twelve-digit national identification number (similar to the Social Security number in the United States). The data was reported to be available for sale by the hackers.

In 2017, Equifax, one of the three large credit bureaus in the United States, was forced to reveal that it had been hacked so thoroughly that it had lost copies of the personally identifiable information for over 145 million people. Information lost included names, Social Security numbers, birth dates, telephone numbers, and addresses. For some people, the lost information included driver's license numbers and credit card numbers. Investigations of Equifax, including by the U.S. Senate, found the company was not following standard policies developed by the information security industry for maintaining secure systems. The company had failed to follow its own internal policies for updating software and auditing its own security. The hacker used a known vulnerability to the Apache Struts product, an open-source system that supported Java-based web applications. Further investigation revealed the hackers had broken in and had access to the Equifax servers for seventy-eight days before being accidentally discovered. Equifax waited a further six weeks after the discovery before making a public announcement.

Credit bureaus, or consumer reporting companies, were private corporations that served an important role in American business and had been around in some form since the mid-nineteenth century. By collecting information on people, such as whether they pay their bills, how much credit they have, and other economic indicators, businesses could determine whether customers were good credit risks. Essentially, such companies made it possible to estimate the risk of a borrower repaying their loan. Such intimate knowledge could be damaging in the wrong hands. The people whose information was in these company databases were not customers but the product; the actual customers were businesses who paid fees to access the credit reporting databases and obtain credit scores for their own customers who they planned to extend credit to.

Logs showed that the credit bureau hackers came from an IP in China. There was no evidence that any of the data stolen from Equifax had appeared on any dark web cybermarket or had been used for identity theft, thus reinforcing the conclusion that this data had been stolen by a nation-state for use in intelligence work, probably by China. A plausible scenario would have been to use the data to identify individuals who have access to classified information or industrial secrets, look up their credit information, and then use financial incentives to compromise those individuals.

## STUXNET

The most dramatic case of documented cyberwarfare came to light in 2010 and became known as *Stuxnet.* Some background is necessary to understand the geopolitical setting of this episode. The United States introduced the Atomic Age when the first atomic bomb was exploded at the Trinity site at White Sands, New Mexico, on the early morning of July 16, 1945. Two atomic bombs were later dropped on the Japanese cities of Hiroshima and Nagasaki, contributing to the end of World War II. These bombs were the end result of the top secret Manhattan Project, costing about $2 billion, to design the new bombs and produce purified weapons-grade uranium and plutonium.

Weapons-grade material for the American bombs was made of heavier isotopes of uranium and plutonium, uranium-235 and plutonium-239, where purification meant accumulating sufficient quantities of these unique substances. One of the main techniques to produce that purified material was by using thousands of centrifuges, where the spinning machines separated heavier isotopes from lighter isotopes. Doing this process over and over again accumulated ever more pure amounts. More efficient techniques were later invented to purify the weapons-grade material, and the centrifuge method was retired.

The Soviet Union joined the nuclear club in 1949 by testing its own atomic bomb, which was built using technology purloined by communist spies from the Manhattan Project. The United Kingdom joined the nuclear club in 1952, France joined in 1960, and China joined in 1964. These five nations formed the basic core of nuclear powers, though the two superpowers made the vast majority of weapons. Israel developed its own ability sometime in the 1970s but kept it secret, and there were no known actual tests of an Israeli atomic bomb. India tested an atomic bomb in 1974. South Africa built six bombs of its own (probably assisted by the Israelis), which were dismantled when white-majority rule was ended and Nelson Mandela became president in 1991.

In 1998, Pakistan tested its own atomic bomb, restarting the global nuclear arms race among minor states. The Pakistani nuclear physicist A.

Q. Khan (1936–2021) had built the bomb, returning to the old technique of using thousands of centrifuges to enrich the weapons-grade material. Khan firmly believed that every nation had the right to possess its own atomic weapons, and he transferred technology and sold expertise to a variety of nations, including North Korea, Libya, and Iran. In 2006, North Korea joined the nuclear club, while Iran worked vigorously on its own program to build atomic weapons, even though it publicly denied that this was the goal of its program. Iran claimed to only be enriching uranium and plutonium for peaceful uses in civilian nuclear power plants, though weapons-grade material is too refined for use in civilian reactors.

Now entered Stuxnet. The malware was first discovered by a small Belarusian antivirus company in 2010. Antivirus companies designed their products to report back to their company servers from their customers' machines with files that were suspicious. This was how such companies found new malware, and such reports also allowed them to gauge how far particular flavors of malware had spread. Such companies also shared their finds with other antivirus companies around the world so that all the companies could react quickly to new threats. The new malware was very sophisticated, only 500 kilobytes in size, and contained four zero-day attacks, an attack that can occur quickly after a weakness in software was discovered, but no one could figure out what this malware did.

A Russian antivirus company, Kaspersky Lab, eventually figured out the purpose of Stuxnet. It sought out specific types of computer-controlled centrifuges sold by the German conglomerate Siemens. The malware then infected the centrifuge's computer and caused it to misbehave by sending out erroneous information and ran the machines so quickly that they physically damaged themselves. Iran had purchased over 8,000 Siemens centrifuges for its uranium enrichment program, and reports quickly emerged that it had been suffering significant problems with the centrifuges for the past year.

Stuxnet was not intended to circulate in the wild on the regular internet outside of the Iranian processing facilities. It was probably introduced into the Iranian nuclear weapons program via an infected USB drive, possibly by a human spy in the facility, and then had spread through the local networks into the centrifuges. Stuxnet acted as a worm in that it analyzed its local network environment and then transmitted copies of itself to other computers that connected to that network. Once the malware escaped, it was only a matter of time before copies of it came to the attention of antivirus companies.

Who made Stuxnet? The obvious choices were Israel, the United States, and other nations who were opposed to Iran's nuclear ambitions. Both Israel and the United States had the capability to make such a tool, suspicions that were confirmed two years later by a series of *New York Times* articles that described the two countries building Stuxnet as part of a cyberwar operation called "Olympic Games." The vice chairman of the Joint Chiefs of Staff, a Marine four-star general named James Cartwright, was later convicted of lying to federal investigators over his role in revealing this information to a reporter. Barack Obama pardoned him during his last days as president.

The designers of Stuxnet also used parts of the same programming code and zero-day exploits to make other malware that was primarily used in the Middle East. The Flame malware was forty times the size of Stuxnet and was used to spy on computers. Later research showed Flame had been released before Stuxnet and was a parent, not a child, in the chain of innovation. Flame spread through multiple means, including USB drives, through Bluetooth network connections, and through the Microsoft Windows update utility. Other malware tools, probably built by the same people, included the malware packages Duqu and Gauss, tools used to gather intelligence from infected machines, including searching through PDF files for keywords and transmitting the files that matched the keywords. Like most sophisticated malware, these tools patiently transmitted their purloined information back to command and control servers in small packets of data spread across long stretches of time, which made it harder for network monitoring programs to notice them.

Iranian hacking groups became active on their own, most likely sponsored by their military and government. In 2012, the computers of Saudi Aramco, Saudi Arabia's state petroleum company, were attacked by an Iranian group calling itself Cutting Sword of Justice. Using a phishing email to gain access, the hackers used wiping software called Shamoon on 35,000 Aramco computers, leaving behind empty hard drives. It took months for the company to recover. This was an example of cyberwar, though it did not lead to a physical shooting war. Iranian cyberattacks were also made on Western banks from 2011 and 2013. In 2020, an Iranian group known as APT 35 or Charming Kitten, launched unsuccessful efforts to penetrate American electric utilities and oil and gas companies. They also made unsuccessful phishing efforts to penetrate President Trump's 2020 reelection campaign.

## THE PATRIOT ACT AND EDWARD SNOWDEN

When terrorists attacked the United States on September 11, 2011–also called 9/11 because the number matched the three-digit phone number used for emergency telephone calls in the United States—almost 3,000 people died. Americans quickly reverted to a Cold War mentality, where the federal government concentrated on ensuring national security, and there was widespread support in Congress and among the American people for all efforts to prevent a repeat of that awful day. Congress quickly moved to pass the Patriot Act (USA PATRIOT awkwardly meant Uniting and Strengthening America by Providing Appropriate Tools Required to Intercept and Obstruct Terrorism). Federal law enforcement was given extra powers, including expanding the power for the FBI to issue National Security Letters, which were effectively search warrants where a judge was not consulted. Such a letter could be used to compel internet service providers (ISP) to not only turn over a person's web browsing history but also forbid the provider from informing the subject of the letter that they were being investigated.

The Bush administration decided that its measure of success in its counterterrorism efforts after 9/11 would be no more terror incidents on American soil. Members of the administration also subscribed to the strong unitary executive theory of presidential authority, which meant that the clause in the U.S. Constitution that made the president commander in chief of the armed forces gave the president virtually unlimited powers. Using this theory, the administration secretly ordered the NSA to begin mass surveillance of American citizens.

This went beyond the generous provisions of the Patriot Act. There already was an established mechanism to provide surveillance of individual Americans or other people within the United States: FISA warrants. The Foreign Intelligence Surveillance Act (FISA) created a secret court to issue search warrants that permitted surveillance of Americans within the United States or abroad and foreigners in the United States. Non-Americans outside the boundaries of the United States were not protected by American law and were readily surveilled by the NSA as part of its regular mission. The secret FISA court was created in 1978 in reaction to the revelations in the mid-1970s that the FBI, CIA, and NSA had engaged in illegal domestic surveillance. The Bush administration found the FISA process too cumbersome for its taste and believed that the president could supersede the FISA restrictions under its interpretation of unitary executive power, but the surveillance was kept secret because the administration knew that most constitutional scholars disagreed with it.

Among other efforts, the Bush administration issued secret orders to the NSA to start to accumulate metadata on all phone calls in the United States. *Metadata* is data about data; in the context of telephone calls, metadata is what phone number made a call, what phone number received the call, how long the call lasted, and the location data for where the call was made from and received. The content of the calls, a recording of the calls, would be the data of the call, not the metadata. Telecommunications companies regularly kept metadata on calls as part of their normal business so that they could accurately bill customers for usage and to help diagnose system problems. The NSA requested that the telecommunications companies turn over copies of all their cell phone metadata. The telecommunications companies, which are heavily regulated and dependent on federal goodwill, did not object, even if they thought that the requests were illegal. The NSA and law enforcement agencies often found it easier to obtain data on American citizens by asking corporations for the data rather than obtaining the data through their own surveillance capabilities.

An NSA contractor named Edward Snowden (1983–) gained notoriety in 2013 by exposing the NSA's operations to the public. Snowden was one of the hundreds of thousands of private contractors who worked for the federal government, a situation particularly common in the computer field because private contractors were a quick way to increase the skilled workforce without the obligation of keeping those employees for a long period of time. As a contractor for the NSA, Snowden had been cleared for top secret access. He worked as a system administrator, which gave him access to lots of files and data. Snowden was appalled by the movement of the NSA from targeted surveillance of individuals to mass surveillance of Americans. He decided to change from working for the government to working for the public interest.

Snowden copied a large amount of files and data from the NSA servers, encrypted them, and prepared to reveal what he had learned. He contacted journalists from the United Kingdom, communicating through encrypted emails over the encrypted Tor network, and he arranged to meet with the journalists in a hotel in Hong Kong. He was deeply committed to his mission, articulate in his views, and realized that he was not going to be able to return to a normal life. After the federal government realized that its NSA secrets were loose, it tried to regain control of Snowden, but the whistleblower fled to Moscow, where Russia gave him asylum. As of 2022, he still lived there with his American girlfriend, who originally had no idea what he was up to. One of the reasons that Snowden refused to return to the United States was because he believed that he would not be given a fair trial because he could not use being a whistleblower as a defense. Such a defense was restricted by case law. Snowden never released all the information he had copied from the NSA, keeping key documents as insurance. To many people, Snowden was a hero, worthy of a movie about his life and a possible candidate for a presidential pardon. To other people, he was a traitor whose place of refuge in Russia was a source of deep irony.

The first big revelation that the press released from the Snowden documents was that the NSA was stockpiling telephone metadata. Other Snowden documents showed how effective the NSA was at vacuuming up data from the rest of the world in cooperation with the other members of the Five Eyes. The NSA later abandoned the telephone metadata project. The Patriot Act was extended multiple times by Congress, with only slight restrictions added. One of the extensions exempted telecommunications companies from legal liability for complying with past illegal requests from the NSA.

## 2016 ELECTION

New technologies had always affected the electoral process. Early elections in the United States used to take weeks to resolve, as ballots had to be counted by hand and the results collected by postal letters. By the latter half of the twentieth century, American elections were usually called within hours of the closing of the polls. Computers were a big reason for the changes in elections, as demonstrated in the earlier account about the UNIVAC and the 1952 presidential election, and that success relied on a relatively simple statistical model based on data from previous elections. Political campaigns learned to use computers to build more sophisticated models of how voters acted based on records of earlier votes. This led to more effective campaigning and also increased the ability of political parties to gerrymander voting districts. Gerrymandering had existed for a long time; state legislatures routinely drew lines for congressional voting districts and local voting districts that they thought would benefit their own parties. Such guidelines were often used to isolate African American voters into a smaller number of districts. By the 1980s, as more data became available, gerrymandering politicians had started drawing districts that moved the boundaries based on neighborhoods and even streets by isolating the precise demographics and how such neighborhoods had voted in the past.

Every new innovation in communication technologies had also been adopted by politicians in their efforts to get votes: a technological journey beginning with pamphlets and proceeding through newspapers, radio, television, internet advertising, and finally social media. The innovations also benefited fundraising. Howard Dean (1948–), the former governor of Vermont, ran in 2003 for the Democratic nomination for president. As a dark horse candidate, he attracted little attention until people noticed how much money he was raising for his campaign. Unlike most other politicians, who sought out donors who could donate the maximum allowable amount, Dean's campaign developed email lists of small donation donors who the campaign would regularly tap for amounts less than $100. In the past, this kind of effort would have been too much work to create and sustain. But email was cheap, and the internet had become mainstream as a way to communicate. Not a lot of people could afford to donate thousands of dollars, but much smaller amounts were more manageable; donors also showed a commitment that could later be used to form grassroots networks of campaign volunteers. Though Dean flamed out early in the primary season, he had shown the way to the future.

A veteran of the Dean campaign became the chief digital strategist for the 2012 Obama reelection campaign, and Obama's staff showed similar skills in raising impressive amounts of small donations from a larger base of supporters. The Obama campaign also showed the power of big data in 2012. Obtaining vast amounts of demographic data from the U.S. Census Bureau, credit bureaus, mail lists, voting lists, and the like, the Obama campaign went from just profiling groups of voters to identifying individual voters. Up to a thousand variables were used to identify each targeted voter. Based on their demographics, statistical models were used to calculate the probability of individual voters actually casting a vote and the probability of voting for President Obama.

The Obama campaign had been building this microtargeting model even before the previous presidential campaign, and it believed that it could literally identify every voter by name who had voted for Obama in 2008, even though the actual vote cast remained secret. The campaign just needed to convince those voters by individual outreach to vote for Obama again. The "get out the vote" effort was very important to the campaign because usually only about six out of every ten eligible voters actually vote in presidential elections. Polling had shown which states were battleground states, and getting the right voters out to vote could swing that state to the Democrats. The Republican campaign had a similar effort and also used sophisticated data mining, but it did not do as good of a job. For that and other reasons, Obama handily won the 2012 election. Running a political campaign had moved far beyond voter demographics, and in its own way, data modeling empowered the individual voter, at least in swing states.

The 2016 presidential election in the United States was the first election in which foreign interference became a serious issue. With its worldwide reach, the internet made it easy for foreign actors to access American voters from afar, and social media proved to be the ideal mechanism to reach out to voters. Because Americans would have reacted poorly to explicit efforts by foreign actors to influence the election, such foreign actors had to disguise their efforts to use social media sites to spread misinformation. While China, Iran, and other nations have made such efforts, the primary actor in the 2016 election was Russia. These efforts were well documented after the election in declassified reports from the American intelligence community, the U.S. Senate, and a Department of Justice investigation (commonly called the Mueller Report, after the independent counsel who directed the investigation). Considerable partisan rancor over these reports rarely confronted the facts as presented.

The Russian effort was founded on the well-established ideas of propaganda and the prior Soviet practice of "active measures," which were efforts to influence public opinion, often through deceit (as noted earlier, PsyOps). The Russians wanted to "sow distrust and discord and lack of confidence in the voting process and the democratic process." Much of the Russian effort supported the Trump campaign and opposed the Clinton campaign, perhaps a reflection of the personal animosity that Russia's leader, Vladimir Putin, felt for Hillary Clinton; even so, the main goal of the Russians was to spread confusion and erode trust. Why would Putin not want Americans to trust each other? Putin, an old KGB hand, operated under the understanding that trust undermined dictatorships because it allowed relationships to form on a basis other than fear. Trust would make a society stronger, and a strong America raised difficulties for Russia's efforts to regain control and influence in the neighboring countries of Ukraine, Georgia, and the Baltic states.

The Russian effort to interfere in the 2016 election had four prongs. The first, an influence effort, included setting up fake accounts on Facebook, Instagram, Twitter, YouTube, Tumblr, LinkedIn, and other social media sites so that their posts looked like they came from fellow Americans. They also bought Facebook and other types of advertisements and used their network of fake accounts to "like" each other, creating a sense of groundswell of sentiment on social media sites. Examples of Russian ads included a picture of a smiling Clinton with a Muslim woman and the words "Support Hillary. Save American Muslims." written in a font that reminded the reader of traditional Arabic. Another ad had a picture of Jesus arm wrestling with Satan, with the following text (in CAPS):

SATAN: IF I WIN CLINTON WINS!
JESUS: NOT IF I CAN HELP IT!
PRESS "LIKE" TO HELP JESUS WIN!

A private company in St. Petersburg owned by a Russian oligarch, the Internet Research Agency (IRA), ran many of these fake accounts and acted as a "troll farm." A 2015 report found that the IRA had an estimated 400 employees working as trolls on twelve-hour shifts, doing their best to throw sand into the gears of internet communication. The employees had maintained groups of fake online personas, called "sockpuppets," to build up online reputations and relationships to help them in their deceptions. IRA had earlier run dozens of such operations in support of Russian subversion in Ukraine and other areas of Russian interest. State-funded Russian media also produced overt content to support the influence effort, including the Russian government-owned cable station RT America TV. The whole effort was designed to be deniable by using front organizations, agents of influence, and false flag operations. The Russians also used bots, programs that were designed to act as users, in an effort to amplify their impact. How effective this effort was remains difficult to gauge, but post-election analysis of Facebook records showed that Russian troll activity did have an amplifying effect on conspiracy theories. The Russians sought to amplify messages from the far left and far right of the American political spectrum, drowning out the "majority whispers" of the political center. In a celebration on the night of the 2016 election, IRA employees "uttered almost in unison: 'We made America great.'"

In the second prong, the GRU (Russian military intelligence) also conducted influence operations through social media, but in addition, it hacked into the Democratic National Committee (DNC). GRU officers sent hundreds of spear phishing emails to various people in the Clinton presidential campaign in March 2016. They were able to obtain access to various email accounts and the DNC network, allowing them to place two types of monitoring software on DNC servers in April, called X-Agent and X-Tunnel. The first program was a powerful tool that vacuumed up data from servers and PCs, logged keystrokes, and took screenshots. Logging keystrokes was an excellent way to grab usernames, passwords, or messages—anything that users typed into their machines. The second tool created an encrypted connection through the DNC network defenses to exfiltrate out the purloined data. These tools were controlled by a program on a leased computer in Arizona.

The third prong of the Russian effort was GRU's hacking attempts to break into the computerized infrastructure of the American voting system. This included efforts to access voter registration databases at the state level and the actual voting machines and networks. While this GRU effort was a concern, the Senate investigation found no evidence that the actual voting process was compromised on Election Day. The first known incident came when Russian hackers penetrated the Illinois state voter registration system in June 2016. The hack used an SQL attack that was not initially detected by Illinois employees, but three weeks later, they did notice heavy loads on the database servers as the hackers ran database queries to access information, leading the system administrators back to a log of the original penetration. Successful hacks and unsuccessful attempts were also made on other state systems. Phishing attacks were common in this effort because they were so easy and effective.

The fourth prong of the operation was to distribute the purloined DNC information in a way that would damage the Clinton campaign. The GRU used a service that concealed the identity of the registrant to reserve the web domain DCLeaks.com just a week after penetrating the DNC. Then it started posting stolen documents on the site taken from compromised personal email accounts in June 2016. Russian agents also operated a Facebook page for DCLeaks to communicate with interested journalists.

The DNC announced in June that it had been hacked, and an analysis by the security firm they had hired concluded that it was by Russian state-sponsored hackers who formed a group they called Fancy Bear. The GRU immediately created a false story to divert attention, creating an online persona called "Guccifer 2.0" to create a blog claiming that the DNC server hack was really the work of a single individual from Romania. This blog used unique words or phrases from English, such as "some hundred sheets," "illuminati," and "worldwide known." GRU's involvement was made obvious because internet searches for these exact terms had been conducted from a GRU server in Moscow in the two hours previous to Guccifer 2.0's first blog post, as if a GRU officer was making sure that they were using the English terms correctly before making the post. The prize jewels from the DNC, emails and documents showing conflict between the Clinton and

Bernie Sanders branches of the Democratic party, were fed by the GRU to WikiLeaks to be released in the weeks before the election.

WikiLeaks had been founded in 2006 by Julian Assange (1971–), an Australian hacker and political activist, as a location to upload and store all types of secret documents in an effort to make the world a more transparent place. The site was designed to be distributed across many servers so that it would be difficult to knock the site off-line with DDoS attacks or because it ran afoul of an individual nation's laws. WikiLeaks served up millions of documents on many topics. This radical form of transparency excited many other activists and antagonized powerful actors. WikiLeaks published internal documents from the Church of Scientology, other private documents that hackers had stolen elsewhere, and classified information from the U.S. government stolen by an enlisted soldier in the army, Bradley E. Manning (1987–), in 2009 (Manning gender transitioned while in prison to Chelsea Manning).

Assange traveled the world as a celebrity activist, raising money to support WikiLeaks. He was arrested in the United Kingdom in 2010 on a Swedish warrant investigating sexual crimes in Sweden. During the course of legal proceedings, Assange sought political asylum in the Ecuadoran embassy in London, where he stayed from 2012 to 2019. During the 2016 presidential campaign, Assange worked with the GRU to obtain the hacked DNC materials and release them. In communications between Assange and the GRU, it was clear that Assange wanted to release the material in a way that would do the most damage to the Clinton campaign. WikiLeaks released over 50,000 documents in total taken from Democratic political consultant John Podesta's personal email account. Assange and WikiLeaks also made public statements that the hacked materials were not from the GRU but came from a disgruntled former DNC staff member who had later been murdered. This story was not true. While the Trump presidential campaign cheered on WikiLeaks and had contacts with the organization, the Mueller investigation did not find a smoking gun showing collusion. However, such collusion or coordination was not necessary to release information from the DNC hacks in ways that maximized damage to the Clinton campaign.

After being evicted from the Ecuadoran embassy in 2019, where he had overstayed his welcome, Assange was sentenced to a British prison for skipping bail and was served with an extradition request from the U.S. government and charged with violating the Espionage Act of 1917. Assange maintained that the Swedish investigation and American espionage indictment were efforts to silence WikiLeaks. The U.S. government had noticed that Assange and WikiLeaks were particularly interested in exposing perceived malfeasance on the part of the United States while taking care to not criticize Russia. In 2017, the CIA director described WikiLeaks as a "a non-state hostile intelligence service," which was "often abetted by state actors like Russia." Assange claimed that he was a journalist and deserved the free speech protections that journalists enjoyed, while the federal government claimed that he was primarily a hacker and did not deserve such free speech protections.

Some supporters of WikiLeaks had become discouraged by what they saw as a site that catered too much to Assange's biases, so they founded a separate effort, a collective called Distributed Denial of Secrets, or DdoSecrets. Like WikiLeaks, many of its most impressive revelations came from hacktivists, though unlike WikiLeaks, its policy was to make clear when the source of its revelations come from state-sponsored hackers. DdoSecrets was also willing to publish data detrimental to Russian interests, such as the 175-gigabyte trove it called "The Dark Side of the Kremlin" in 2019, three years after it had been hacked out of Russian servers and WikiLeaks had declined to publish it.

Despite indications that Russia was meddling in the election, the Obama administration chose to not publicly discuss this issue in 2016. It did not want to be seen as trying to influence the outcome of the election. In the end, the Mueller investigation led to federal convictions of Trump's campaign manager, Trump's personal lawyer, and the political consultant that had the closest contacts with WikiLeaks. Several other aides were also convicted. These convictions were mostly based on the charge of lying to federal investigators. Mueller indicted thirty-four people, most of them Russian nationals, and three companies, including IRA. These indictments of foreigners were unlikely to lead to trials unless the individuals were caught in a third country that would extradite them to the United States.

The Russian campaign was an example of asymmetrical warfare, where an adversary recognized they were in a weaker position than their opponent in using traditional levers of power, such as military force or economic dominance, and so resorted to cyberwarfare, guerilla warfare, or ideological struggle. The Senate report declared, "Russia's aptitude for weaponizing internet-based social media platforms against the United States resulted from Moscow's experience conducting online disinformation campaigns against its own citizens for over a decade." Such campaigns will probably become even more common in the future.

A year later, the same GRU hacking group, Fancy Bear or Sandworm, that broke into the DNC servers stole nine gigabytes of data from the French presidential campaign of Emmanuel Macron and then released the emails and documents via social media in an attempt to damage his campaign. The campaign claimed that among the genuine documents were sprinkled fake documents. Macron won anyway. This group was also held responsible for the cyberattacks on the 2018 Winter Olympics in South Korea, an event that Russian athletes had been banned from as punishment for drug doping. The GRU hacking group had a formal name, Unit 74455, though it had accumulated various other names. Information security researchers had given it names before eventually learning the true primary source of the hacking. For instance, the name for Sandworm came from references to the classic science fiction novel by Frank Herbert, *Dune*. In October 2020, the federal government indicted six Russian men who worked for Sandworm and had launched NotPetya and had also hacked the Macron campaign and the 2018 Winter Olympics.

## VAULT 7

The CIA and NSA in the United States were also vigorous practitioners of cyberespionage. The charters of the CIA and NSA effectively required the government of the United States to spy on the rest of the world. How effective had the CIA and NSA been at this activity? An event in 2017 certainly embarrassed them. WikiLeaks started to publish what it called "Vault 7," a trove of CIA hacking tools that had somehow been copied from a CIA division called the Center for Cyber Intelligence (CCI). These hacking tools were later used against the United States.

Among the hacks, including numerous zero-day attacks, was a way of turning a certain model of Samsung smart televisions into remote microphones, allowing a spy to listen to conversations in a room even when the television was not turned on. The CIA and the United Kingdom's MI5 agency had jointly developed this ability and called it "Weeping Angel." The name came from the British science fiction series *Doctor Who*, in which Weeping Angels are a frightening alien species who appear to be statues and only move when you are not looking at them. Other hacks targeted network devices, smartphones, and different operating systems.

A Senate investigation found that the CCI had prioritized its hacking activities while following "woefully lax" security practices themselves. The investigators admitted that if a nation-state had stolen the Vault 7 tools and kept that fact secret, the CIA would not even know it had been comprised. The versions of the hacking tools that were stolen were not always the final version of the tools, implying that the theft occurred from a development server environment. The CCI was a sloppy spy agency. Sensitive cyberweapons were not compartmentalized from each other, meaning that if a hacker penetrated one cyberweapon program, it was easy to get into other cyberweapon environments; users shared system administrator passwords; and removable media controls were not effective, meaning that a user could copy data onto CD-ROMs or USB drives. All of these practices were banned in other classified systems by national policy, but the American intelligence agencies had been exempted from these good security practices because they were supposed to know better. Because WikiLeaks has apparently not released all of the data found in the Vault 7 trove, the Senate investigation could only determine that somewhere between 180 gigabytes and 34 terabytes was stolen.

The NSA had a similar data loss a year before the CIA was hacked. A group called Shadow Brokers appeared in 2016 and claimed to have hacked into the NSA hacking group, Tailored Access Operations, and stolen the NSA hacking tools. The name Shadow Brokers may have been copied from the game *Mass Effect*. The group tried to auction off what they had stolen, and samples from their trove were posted on the Web, which showed that the group actually had what it claimed to have. The auction accumulated less than $1,000 in bitcoins. Snowden was of the opinion that the group was not interested in the auction but were really state-sponsored Russian hackers who wanted to show that the NSA had been hacking the world. The Russians had just recently been accused of hacking the DNC, and perhaps they wanted to show that the United States engaged in similar activities.

It is ironic that Snowden would advance this theory, especially since he relied on the protection of the Russians and because he had been responsible for the NSA's other greatest breach. As damaging as Snowden's revelations had been, they had not included technical details of hacking tools. The Shadow Brokers' revelations were much more serious because they included the actual tools. The Shadow Brokers gradually trickled out more information over the later months. They released about twenty NSA hacking tools, some of which contained zero-day exploits, in April 2017.

One of these exploits was called EternalBlue, a flaw in Microsoft's Server Message Block (SMB) networking protocol, where specially crafted packets could maliciously run code on the target machine. Quietly alerted by the NSA a month before the Shadow Brokers released the source code for EternalBlue, Microsoft had released a patch to fix the problem. Only a month after the Shadow Brokers leaked EternalBlue, a new ransomware attack called WannaCry appeared. Many major sites had not yet upgraded their Microsoft operating systems with the new patch, a common situation because large IT organizations check all patches before implementing them because they do not want the patches to accidentally break any of their systems. WannaCry hit like a cyberstorm, infecting major systems around the world, including Britain's National Health Service, Spain's Telefónica, the German railway company Deutsche Bahn, and France's Renault. Security researchers named the malware WannaCry because the malware added the file extension .wncry to files it had encrypted. The attackers encrypted files and demanded ransoms.

WannaCry had a clever feature; the malware checked for the existence of a website before it finished infecting the machine. The website did not actually exist, which was a signal for WannaCry to finish infecting the machine. This was an effort to prevent WannaCry from running in a virtual environment. Security researchers often obtained a copy of malware and then ran it in a virtual environment, where software simulated the hardware and ran a copy of the operating system. The malware would have no clue that it was running in a software virtual machine and not on real hardware. This allowed the researcher to watch the malware and see what it did and, if necessary, to run the malware over and over again in fresh virtual machines. The programmers of WannaCry wanted to thwart such research efforts, so the malware looked for a website that did not exist, based on the assumption that a virtual machine would pretend that such a website existed.

Marcus Hutchins (1994–), a twenty-two-year-old British security researcher, had been looking at the code of WannaCry when he ran across the website name in the code. He looked the site up and found that it was not registered, so it did not return a value. He was not sure what the site was going to be used for, but the site might be a future command and control node for the malware. So he registered the website. Suddenly, all around the world, WannaCry stopped infecting new machines. Hutchins was the hero of the moment, though he did not like the accolades. He was later arrested during a trip to the United States and pleaded guilty to writing malware, including the Kronos malware, when he was a teenager and young adult, before he became a white hat hacker. The judge considered his service in stopping WannaCry; he was sentenced to time served and probation.

After the WannaCry attack, the president of Microsoft wrote on the company's official blog that the company had 3,500 security engineers, a Microsoft Threat Intelligence Center, and a Digital Crimes Unit, all trying to stop such events. He decried the stockpiling of zero-day digital vulnerabilities by governments instead of working together with the industry to solve these problems. He characterized the recent losses of such secrets by the CIA and NSA as being equivalent in the physical world to the American military having some of its Tomahawk cruise missiles stolen. It was a limited comparison. Cyberattacks could cause a lot more damage, of a different kind, than a few cruise missiles with conventional warheads. In 2020, the NSA did alert Microsoft to a significant flaw in its encryption software, perhaps a harbinger of a more cooperative approach by American intelligence agencies. The NSA's founding charter did require it to both be a spy and to defend American "communications intelligence" systems.

Various strands of evidence pointed to the origin of WannaCry as the Lazarus Group. These hackers were also responsible for the previous attack on Sony Pictures in 2014 and online theft from the central bank of Bangladesh in 2016. They were thought to be a North Korean government hacking organization. WannaCry was odd in that it was mostly a proof-of-concept attack, and the system to receive bitcoin payments was so sloppy that the few companies who paid the ransom did not receive functional decryption keys.

A month after WannaCry, Sandworm launched a similar attack called NotPetya. This was only the latest in a series of attacks aimed at Ukraine, a nation that had been part of the Soviet Union until 1991, when it gained its independence with the breakup of the Soviet Union. Like other former Soviet states, Ukraine had struggled with economic problems, government corruption, and, later, meddling in its internal affairs by a resurgent Russia. Ukraine had a large Russian minority that was a majority in parts of the country, such as Crimea and parts of eastern Ukraine. After clandestine Russian forces invaded and took Crimea from Ukraine in 2014, the United States and European nations issued sanctions against Russian individuals and companies in punishment. Russia later sent troops and mercenaries into eastern Ukraine, where they combined with Russian Ukrainians in a low-grade war to strip away those parts of Ukraine. Earlier hacking attacks had brought down Ukrainian power networks and Ukrainian companies, and there were also substantial disinformation efforts.

NotPetya crippled substantial parts of Ukraine's information technology infrastructure, which would take weeks to repair, but the malware also spilled over into the rest of the internet, a common problem with such programs. The entire information technology infrastructure of Maersk, the world's largest shipping company, was knocked down: 4,000 servers and 45,000 PCs. It took weeks to recover, and the company estimated that the outage and recovery costs were at least $250 million to $300 million. Maersk management regularly approved requests for security upgrades after this, such as multifactor authentication and quickly moving computers to the latest versions of operating systems. Merck, the pharmaceutical giant, lost 15,000 Windows computers in just ninety seconds to NotPetya. The malware had originated as a ransomware product, but there were few attempts to demand ransoms, as NotPetya just encrypted data and forced IT organizations to reinstall operating systems, programs, and data from backups. NotPetya was often characterized as the worst cyberattack up to that time because estimates of the damage reached $10 billion. When Russia invaded Ukraine in early 2022, the West anticipated a similar attack.

Although many cyberattacks have been described in this chapter, they are only a small fraction of the total number of attacks and other information security incidents since the 1980s. The number of attacks and their ferocity only grew as the twenty-first century marched on.

# Reflections on the Past and Future: A Conclusion

In our explorations of computing up to this point, we have seen many factors influencing the course of innovation: technical standards or the lack of them, market dominance, intellectual property rights, technological momentum, paths not taken, public interest, externalities, costs and difficulties in manufacturing, engineering needs, personal drives and world views, releasing premature products too far ahead of the technological or business curve, political considerations, failure or success in positioning for technological or market changes, and, perhaps surprising to some, the influence of science fiction.

Our descriptions of technologies and scientific advances barely scratch the surface of the possible futures created by advances in computing. While narratives of progress are often regarded with suspicion by historians and other scholars, few will argue that the story of the rise of information machines is a story of computing technology growing ever more sophisticated and ubiquitous. The computer has opened up knowledge and global communications to multitudes, just as the printing press did. By improving the means by which we communicate, entertain ourselves, travel, calculate, and do millions of other tasks, the computer could be an essential tool in reaching for the stars and our quest for all that might follow; including better understanding ourselves.

However, the computer has accelerated the pace of technological change so much that it sometimes becomes difficult to imagine our place in the future. Science fiction may imagine these things. With computing, reality is not far behind, and it is even often ahead of our imagination. Current research and development with computing promises even more tantalizing

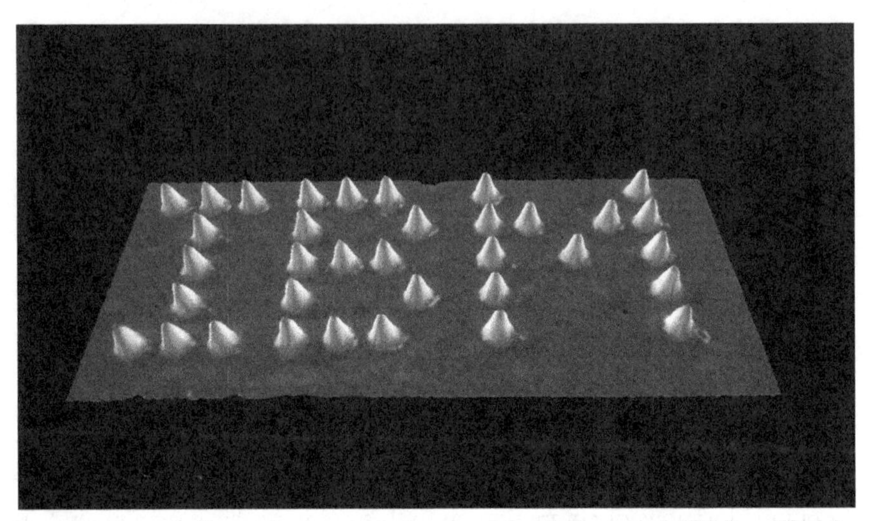

Scanning tunneling microscope photograph of the word IBM spelled in xenon atoms. IBM Corporate Archives. (Reprint Courtesy of IBM Corporation ©)

and scary science fiction futures, both obvious (such as quantum computing, nanotechnology, encoding DNA with information, or artificial intelligence [AI]) and those scarcely imagined. A reader fifty years from now will likely look back on the computers and software available in the first quarter of the twenty-first century and be astonished at how primitive it all appears.

# Bibliography

Abbate, Janet. *Inventing the Internet*. Cambridge, MA: MIT Press, 2000.

Ammirati, Sean. *The Science of Growth: How Facebook Beat Friendster—and How Nine Other Startups Left the Rest in the Dust*. New York: St. Martin's Press, 2016.

Barrat, James. *Artificial Intelligence and the End of the Human Era: Our Final Invention*. New York: Thomas Dunne Books, 2013.

Bashe, Charles J., Lyle R. Johnson, John H. Palmer, and Emerson W. Pugh. *IBM's Early Computers*. Cambridge, MA: MIT Press, 1986.

Berlin, Leslie. *Troublemakers: Silicon Valley's Coming of Age*. New York: Simon & Schuster, 2017.

Berners-Lee, Tim. *Weaving the Web: The Original Design and Ultimate Destiny of the World Wide Web by Its Inventor*. San Francisco, CA: HarperSanFrancisco, 1999.

Brunton, Finn. *Spam: A Shadow History of the Internet*. Cambridge, MA: MIT Press, 2013.

Budiansky, Stephen. *Code Warriors: NSA's Codebreakers and the Secret Intelligence War against the Soviet Union*. New York: Alfred A. Knopf, 2016.

Burks, Alice R., and Arthur W. Burks. *The First Electronic Computer: The Atanasoff Story*. Ann Arbor: University of Michigan Press, 1988.

Campbell-Kelly, Martin. *From Airline Reservation to Sonic the Hedgehog: A History of the Software Industry*. Cambridge, MA: MIT Press, 2003.

Campbell-Kelly, Martin, William Aspray, Nathan Ensmenger, and Jeffrey R. Yost. *Computer: A History of the Information Machine*. 3rd ed. Boulder, CO: Westview Press, 2014.

Campbell-Kelly, Martin, and Daniel D. Garcia-Swartz. *From Mainframes to Smartphones: A History of the International Computer Industry*. Cambridge, MA: Harvard University Press, 2015.

Carlin, John P., and Garrett M. Graff. *Dawn of the Code War: America's Battle against Russia, China, and the Rising Global Cyber Threat*. New York: PublicAffairs, 2018.

Ceruzzi, Paul E. *A History of Modern Computing*. Cambridge, MA: MIT Press, 1998.

Chandler, Alfred D., Jr. *Inventing the Electronic Century: The Epic Story of the Consumer Electronics and Computer Industries*. New York: Free Press, 2001.

Corn, Joseph. J. *User Unfriendly: Consumer Struggles with Personal Technologies, from Clocks and Sewing Machines to Cars and Computers*. Baltimore, MD: Johns Hopkins University Press, 2011.

Cringely, Robert X. *Accidental Empires: How the Boys of Silicon Valley Make Their Millions, Battle Foreign Competition, and Still Can't Get a Date*. Reading, MA: Addison Wesley, 1992.

Dodge, Martin, Rob Kitchin, and Chris Perkins, eds. *The Map Reader: Theories of Mapping Practice and Cartographic Representation*. Oxford: Wiley, 2011.

Ferro, David L., and Eric G. Swedin, eds. *Science Fiction and Computing: Essays in Interlinked Domains*. Jefferson, NC: McFarland & Company, 2011.

Freiberger, Paul, and Michael Swaine. *Fire in the Valley: The Making of the Personal Computer*. 2nd ed. New York: McGraw-Hill, 2000.

Greenberg, Andy. *Sandworm: A New Era of Cyberwar and the Hunt for the Kremlin's Most Dangerous Hackers*. New York: Doubleday, 2019.

Greenberg, Joshua. *From Betamax to Blockbuster: Video Stores and the Invention of Movies on Video*. Cambridge, MA: MIT Press, 2008.

Hafner, Katie, and Matthew Lyon. *Where Wizards Stay Up Late: The Origins of the Internet*. New York: Simon & Schuster, 1996.

Haigh, Thomas, Mark Priestley, and Crispin Rope. *ENIAC in Action: Making and Remaking the Modern Computer*. Cambridge, MA: MIT Press, 2016.

Hanson, Dirk. *The New Alchemists*. Boston, MA: Little, Brown and Company, 1982.

Hicks, Marie. *Programmed Inequality: How Britain Discarded Women Technologists and Lost Its Edge in Computing*. Cambridge, MA: MIT Press, 2017.

Hiltzik, Michael A. *Dealers of Lightning: Xerox PARC and the Dawn of the Computer Age*. New York: Harper Business, 1999.

Hinsley, Francis H., and Alan Stripp, eds. *Codebreakers: The Inside Story of Bletchley Park*. Oxford: Oxford University Press, 1993.

Ifrah, Georges. *The Universal History of Computing: From the Abacus to the Quantum Computer.* New York: Wiley & Sons, 2001.

Isaacson, Walter. *Steve Jobs*. New York: Simon & Schuster, 2011.

Kaplan, Fred. *Dark Territory: The Secret History of Cyber War.* New York: Simon & Schuster, 2016.

Kent, Steven L. *The Ultimate History of Video Games: From Pong to Pokémon—The Story behind the Craze That Touched Our Lives and Changed the World*. New York: Three Rivers Press, 2001.

Kilday, Bill. *Never Lost Again: The Google Mapping Revolution That Sparked New Industries and Augmented Our Reality*. New York: Harper Business, 2018.

Kline, Ronald R. *The Cybernetics Moment: Or Why We Call Our Age the Information Age*. Baltimore, MD: Johns Hopkins University Press, 2015.

Kushner, David. *The Players Ball: A Genius, a Con Man, and the Secret History of the Internet's Rise*. New York: Simon & Schuster, 2019.

Kushner, David. "The Real Story of Stuxnet." *IEEE Spectrum* (February 26, 2013). Available at https://spectrum.ieee.org/telecom/security/the-real-story-of-stuxnet.

Levy, Steven. *Facebook: The Inside Story*. New York: Blue Rider Press, 2020.

Levy, Steven. *In the Plex: How Google Thinks, Works, and Shapes Our Lives*. New York: Simon & Schuster, 2011.

Lih, Andrew. *The Wikipedia Revolution: How a Bunch of Nobodies Created the World's Greatest Encyclopedia*. New York: Hyperion, 2009.

Macrae, Norman. *John von Neumann*. New York: Pantheon Books, 1992.

Mahoney, Michael. S. *Histories of Computing*. Edited by Thomas Haigh. Cambridge, MA: Harvard University Press, 2011.

Malone, Thomas W. *Superminds: The Surprising Power of People and Computers Thinking Together.* New York: Little, Brown, and Company, 2018.

McCartney, Scott. *ENIAC: The Triumphs and Tragedies of the World's First Computer.* New York: Walker and Company, 1999.

McCullough, Brian. *How the Internet Happened: From Netscape to the iPhone*. New York: Liveright Publishing, 2018.

McNish, Jacquie, and Sean Silcoff. *Losing the Signal: The Untold Story behind the Extraordinary Rise and Spectacular Fall of BlackBerry*. New York: Flatiron Book, 2015.

Menn, Joseph. *Cult of the Dead Cow: How the Original Hacking Supergroup Might Just Save the World*. New York: PublicAffairs, 2019.

Mindell, David. A. *Between Human and Machine*. Cambridge, MA: MIT Press, 2002.

Mindell, David. A. *Digital Apollo: Human and Machine in Spaceflight*. Cambridge, MA: MIT Press, 2008.

Mitchell, Melanie. *Artificial Intelligence: A Guide for Thinking Humans*. New York: Farrar, Straus and Giroux, 2019.

Mollenhoff, Clark R. *Atanasoff: Forgotten Father of the Computer*. Ames, IA: Iowa State University Press, 1988.

Mueller, Milton L. *Ruling the Root: Internet Governance and the Taming of Cyberspace*. Cambridge, MA: MIT Press, 2002.

Nicholson, Matt. *When Computing Got Personal: A History of the Desktop Computer*. Bristol, UK: Matt Publishing, 2014.

Nocks, Lisa. *The Robot: The Life Story of a Technology*. Westport, CT: Greenwood, 2007.

O'Mara, Margaret. *The Code: Silicon Valley and the Remaking of America*. New York: Penguin Press, 2019.

Price, David A. *The Pixar Touch: The Making of a Company*. New York: Alfred A. Knopf, 2008.

Queisser, Hans. *The Conquest of the Microchip: Science and Business in the Silicon Age*. Cambridge, MA: Harvard University Press, 1988.

Raymond, Eric S. *The Cathedral and the Bazaar: Musings on Linux and Open Source by an Accidental Revolutionary*. Rev. ed. Cambridge, MA: O'Reilly & Associates, 2001.

Redmond, Kent C., and Thomas M. Smith. *From Whirlwind to MITRE: The R&D Story of the SAGE Air Defense Computer*. Cambridge, MA: MIT Press, 2000.

Reid, T. R. *The Chip: How Two Americans Invented the Microchip and Launched a Revolution*. New York: Random House, 2001.

Rid, Thomas. *Rise of the Machines: A Cybernetic History*. New York: W. W. Norton, 2016.

Riordan, Michael, and Lillian Hoddeson. *Crystal Fire: The Birth of the Information Age*. New York: W. W. Norton, 1997.

Rojas, Raúl, and Ulf Hashagen. *The First Computers: History and Architectures*. Cambridge, MA: MIT Press, 2000.

Sanger, David E. *The Perfect Weapon: War, Sabotage, and Fear in the Cyber Age*. New York: Crown 2018.

Scharre, Paul. *Army of None: Autonomous Weapons and the Future of War.* New York: W. W. Norton, 2019.

Schmidt, Eric, and Jonathan Rosenberg. *Google: How Google Works.* New York: Grand Central Publishing, 2014.

Schneier, Bruce. *Click Here to Kill Everybody: Security and Survival in a Hyper-Connected World.* New York: W. W. Norton, 2018.

Schneier, Bruce. *Data and Goliath: The Hidden Battles to Collect Your Data and Control Your World.* New York: W. W. Norton & Company, 2016.

Segaller, Stephen. *Nerds 2.0.1: A Brief History of the Internet.* New York: TV Books, 1988.

Siegfried, Tom. *The Bit and the Pendulum.* New York: John Wiley & Sons, 2000.

Singer, P. W., and Emerson T. Brooking. *LikeWar: The Weaponization of Social Media.* New York: Eamon Dolan, 2018.

Sito, Tom. *Moving Innovation: A History of Computer Animation.* Cambridge, MA: MIT Press, 2013.

Slater, Robert. *Portraits in Silicon.* Cambridge, MA: MIT Press, 1987.

Snowden, Edward J. *Permanent Record.* New York: Metropolitan Books, 2019.

Stern, Nancy. *From ENIAC to UNIVAC: An Appraisal of the Eckert-Mauchly Computers.* Bedford, MA: Digital Press, 1981.

Stone, Brad. *The Everything Store: Jeff Bezos and the Age of Amazon.* Boston, MA: Little, Brown and Company, 2013.

Stross, Randall. *Planet Google: One Company's Audacious Plan to Organize Everything We Know.* New York: Free Press, 2008.

Sumpter, David. *Outnumbered: From Facebook and Google to Fake News and Filter-Bubbles—The Algorithms That Control Our Lives.* London: Bloomsbury Sigma, 2018.

Turing, Alan. "Computing Machinery and Intelligence." *Mind* (October 1950): 433–460.

Ullah, Haroon K. *Digital World War: Islamists, Extremists, and the Fight for Cyber Supremacy.* New Haven, CT: Yale University Press, 2017.

Vance, Ashlee. *Elon Musk: Tesla, SpaceX, and the Quest for a Fantastic Future.* New York: HarperCollins, 2015.

Wilkes, Maurice V. *Memoirs of a Computer Pioneer.* Cambridge, MA: MIT Press, 1985.

Williams, Michael R. *A History of Computing Technology*. 2nd ed. Los Alamitos, CA: IEEE Computer Society Press, 1997.

Woyke, Elizabeth. *The Smartphone: Anatomy of an Industry*. New York: New Press, 2014.

Zetter, Kim. *Countdown to Zero Day: Stuxnet and the Launch of the World's First Digital Weapon*. New York: Crown, 2014.

Zuboff, Shoshana. *The Age of Surveillance Capitalism: The Fight for a Human Future at the New Frontier of Power*. New York: PublicAffairs, 2019.

# Index


Abacus, 7–8, 21, 28
Abrams, Jonathan, 198
Abramson, Norman, 126
Adleman, Leonard, 252
Advanced Micro Devices (AMD), 111
Advanced Research Projects Agency (ARPA), 79, 104–105, 107, 121, 124, 126, 129
Advanced Research Projects Agency Network (ARPANET), 107, 123–126, 127–131, 134
Aiken, Howard Hathaway, 39–40
Alferov, Zhores I., 71
Alibaba, 174–175, 185, 216–217
Allen, Paul, 96–97, 118–119
al-Khowarizmi, 6
AlohaNet, 126–127
Altair, 93–96, 97, 98; Altair bus (s-100 bus), 97, 102, 131–132
Amazon.com, 140, 155, 175, 176–179, 180–182, 183–184, 190, 245, 257; Alexa, 175; Amazon A9, 183; Amazon Echo, 175; Amazon Kindle, 175, 179, 181–182; Amazon Prime, 178; Amazon Web Services (AWS), 179, 183–185
Amdahl, Gene M., 80–81, 83
American Association for the Advancement of Science (AAAS), 30, 41
American National Standards Institute (ANSI), 80, 122, 223–224
American Standard Code for Information Interchange (ASCII), 122, 135, 136, 138, 224
America Online (AOL), 132–133, 135, 139–140, 187, 189–190
Analog devices, 1, 28, 30, 52, 123
Analog Planetarium Projection System, 3
Analytical Engine, 15, 18, 20, 25
Anderson, Thomas, 198
Andreessen, Marc, 139
Antheil, George, 259
Antikythera Device, 1, 2, 4
Antitrust, 47, 86, 90, 118, 140, 161–162
*Antitrust* (movie), 266
Apache, 144, 146, 203, 268

Apple Computer, 98–101, 104, 107–110, 115–116, 146, 151–153, 155, 165–170, 173–176, 245, 247; Apple II, 98–101, 107, 116; Apple Watch, 175–176; App Store, 166–167; iPad, 146, 162, 169–170; iPhone, 146, 152–153, 158, 165–170, 173–174; iPod, 151–152, 166; iTunes, 152–153, 166, 181; Macintosh, 104–109, 110, 115, 117, 146, 152; Newton, 162–163, 165–166, 169; Siri, 173–174
Applied Data Research (ADR), 84–86
Arithmetic Processing Unit (ALU), 18
Arithmometer, 14
Artificial Intelligence (AI), 37, 66–67, 79, 122, 173, 224–229, 232, 234, 241
Assange, Julian, 279–280
Association of Computing Machinery (ACM), 89, 108
Astrolabe, 1–2, 4, 9
Atanasoff, John Vincent, 27–32, 35, 38, 41–42, 48–49
Atanasoff Berry Computer (ABC), 27–32
AT&T, 79–80, 122–123, 143, 159–162, 166, 199, 247; and Internet, 129–130
Atari, 97–98, 99, 102, 110, 113, 132, 226
Authentication, 254–255
Automata, 2
Automatic Computing Engine (ACE), 37, 57
Babbage, Charles, 14–20, 21, 29, 39
Babbage, Major-General Henry P., 20, 39
Babylonian Mathematics, 1–2, 5, 27
Backus, John, 63
Baker, W. R. G. "Doc," 59
Baldwin, Frank S., 14
Ball, Joan, 206
Ballmer, Steve, 102, 146, 185
Baran, Paul, 123–124
Bardeen, John, 55
BASIC, 87–89, 96–97, 101, 103–104, 114, 158
Batch processing, 78–79
Bede, Venerable, 8
Bell, Alexander Graham, 122
Bennett, Susan, 173
Berners-Lee, Tim, 109, 137–139, 215
Berry, Clifford E., 29–30, 32, 42
Bezos, Jeff, 176–179, 181, 183–184
Billings, John S., 22
BINary Automatic Computer (BINAC), 46
Binary (Machine) Code, 62–64, 88
Binary Number System, 29–30, 35, 38, 41, 43
Bitcoin, 254, 262–265, 282, 284
Bitnet, 130
Blackberry, 164–165, 167
Blay, Andre, 154
Bletchley Park, 34–37, 44
Blockchain, 263
Blue Origin, 179
Bluetooth, 158, 166, 175, 223, 271
Bluetooth, Harald, 223

Bolt Beranek and Newman (BBN), 124–125, 128
Bombe, 35–36, 40, 44
Brainerd, Paul, 108
Brattain, Walter H., 55
*Breakout*, 99–100
Bricklin, Dan, 101
Briggs, Henry, 10
Brin, Sergey, 142
Brooks, Frederick Phillips, Jr., 80–82
Browser, 138–141, 145, 164, 166, 187, 192, 215
Bulletin Board Systems (BBS) or Computer Bulletin Board Systems (CBBS), 114, 131–133, 135, 187
Burroughs Adding Machines, 24–25, 28, 48, 56–57, 59, 65
Bush, Vannevar E., 3–4, 28, 39, 106
Bushnell, Nolan, 99, 112–113
Busicom, 91–92

C, 80, 89
C++, 89
Calculating rods, 7
Calculators, 8, 11–12, 14, 28, 38, 41, 45–47, 56–58, 75, 91, 94, 100, 158–159, 163, 170
Canova, Frank, 163
Case, Steve, 132–133
Cathode Ray Tube (CRT), 44, 57
Cellular phones/smartphones, 159–162, 164–167, 243
Census Bureau, U.S., 22–23, 46–47
Central Intelligence Agency (CIA), 248–249, 261, 272, 280, 281–282, 284
Central Processing Unit (CPU), 18, 59–60, 76, 78–79, 88–89, 93, 95, 103, 183–185, 211, 256, 262–263
Cerf, Vint, 128
Cheyer, Adam, 173–175
Christensen, Ward, 131–132
Circuit board, 95, 99
Clark, Jim, 139
Clark, Wesley, 124
Clarke, Arthur C., 171
Clement, Joseph, 16–17
Clerical organizations, 22
Clinton campaign, 276–279
Clinton, Hillary, 276–277
Clinton-Gore White House, 136
Clones, 83, 109–111, 112, 114–115, 117
Cloud, 183–185
Codd, Edgar F., 211
Cohen, Stephen, 207
Colligan, Ed, 162–163, 165–166
Colossus, 36–37, 41, 44
Commodore, 97–98, 110, 132
Common Business-Oriented Language (COBOL), 64, 87
Communications, command, control, and intelligence (C3I), 260
Compaq Computer, 78, 110, 117, 165, 237
Compass, 3, 9
Comptometer, 14
Compuserve, 132–133, 139, 187
Computer games, 112–114, 156–158, 207–209
Conseil Européen pour la Recherche Nucléaire (CERN), 137–139

Control Data Corporation (CDC), 65, 89–90
Control Program for Microcomputers (CP/M), 103–104, 131–132
Cooper, Martin, 160
Craigslist, 212
Cray, Seymour, 89–90
Cray Research, 90
Cromemco, 97
Cryptocurrencies, 262–263, 264
Cunningham, Ward, 192
Cyberattacks, 253, 256–259, 261, 271, 281, 284–285

Data General, 77
Davies, Donald, 123–124
de Acosta, Joseph, 8
de Colmar, Charles Xavier Thomas, 14
de Fermat, Pierre, 12
de Prony, Baron Gaspard, 21
de Solla Price, Derek, 1
de Vaucanson, Jacques, 19
Dean, Howard, 275
Decimal system, 5, 7, 41, 43, 63
Decryption, 34, 252–254, 264, 284
Defense Advanced Research Projects Agency (DARPA), 107, 171–173, 232, 234–236
Dell Computer, 110, 117
DeWolfe, Chris, 198
Difference Engine, 15–19, 20–21
Differential analyzers, 3–4, 28, 39, 106
Diffie, Whitfield, 252
Digital devices, 118, 151, 155–156, 158, 169, 198, 207, 219, 221, 233
Digital divide, 241–245
Digital Equipment Corporation (DEC), 77–78, 104, 112, 126, 225
Digital Research Incorporated (DRI), 103–104
Digital Video (or versatile) Disc (DVD), 155
Digital Video Recorders (DVR), 155
Dijkstra, Edsger W., 88–89
Disk drives, 60, 62, 82, 94–95, 103–104, 108, 110, 116
Disk Operating System (DOS), 103, 104, 107, 109, 114–117, 173
Dividers, 9, 11
Domain Name System (DNS), 130–131, 206, 257
Draper, Charles Stark, 73
Draper, John (Cap'n Crunch), 101, 247
"Dread Pirate Roberts," 254
Driverless cars, 234–237
Drones, 232–234, 243, 260
Dubinsky, Donna, 162–163, 165–166
Dudley, Homer, 171

eBay, 140, 188–190, 238, 256, 267
E-Books, 179–183
Eckert, J. Presper, 31–32, 40–43, 46–49
Electromechanical Calculating Machines, 8, 21, 23, 25, 33–40
Electronic calculator, 75, 91; handheld, 75
Electronic Counter-Countermeasures (ECCM), 259
Electronic Countermeasures (ECM), 259
Electronic Data Systems (EDS), 87

Electronic Delay Storage Automatic Calculator (EDSAC), 43–45
Electronic Discrete Variable Arithmetic Computer (EDVAC), 43, 45
Electronic hobbyists, 93–100, 111–112, 132, 146, 154, 156
Electronic mail (Email), 125, 132, 135, 141, 145, 164, 167, 212, 215, 254, 256, 271, 275
Electronic Numerical Integrator and Computer (ENIAC), 4, 31–32, 38, 40–43, 45–46
Encryption, 32–34, 36, 215, 252–254, 261, 263, 266, 284
Engelbart, Douglas C., 106–108, 115, 137, 173
Enigma encoding machine, 33–36, 261
Erasable programmable read-only memory (EPROM), 91
Ethernet, 105, 127, 128, 133, 169
Euclid, 12
Evans, Robert O., 80–81
Expansion card, 95, 97–98, 100

Facebook, 198–199, 202–205, 214, 243, 245, 276–277
Faggin, Federico, 92, 97
Fairchild Semiconductor, 56, 70–71, 90, 99
Fanning, Shawn, 152
Felt, Dorr E., 14, 17
Fernandez, Bill, 99
Ferranti Atlas, 62
Fibonacci (Leonardo of Pisa), 6
FidoNet, 132
File Transfer Protocol (FTP), 125
Filo, David, 141
Finger calculation, 8–9
Flowers, Tommy, 36
Formula Translation (FORTRAN), 63–64, 72, 88
Forrester, Jay W., 52–53
Foster, John Stuart, Jr., 232
Frankston, Bob, 101
Free Software Foundation (FSF), 143–144
FreeBSD, 144
Friis, Janus, 190

Galilei, Galileo, 9
Gates, Bill, 96–97, 102–103, 109, 118–119, 139–140, 146, 179, 228
Gateway Computer, 110, 117
Genaille, Henri, 11
Genaille-Lucas Rulers, 11
General Data Protection Regulation (GDPR), 214–215
General Electric (GE), 56–57, 59–60, 65, 87–88, 150
Geographic Information Systems (GIS), 209
Gerstner, Louis V., Jr., 116
Global Positioning System (GPS), 158, 209–210, 235, 260
GNU (GNU Not UNIX), 143–144
GNU General Public License (GPL), 143–144
Goldstine, Adele and Herman, 46
Google, 142, 147, 167–168, 175, 179, 182–185, 193, 198, 201–202, 214, 216, 225–226, 229, 236, 243, 245, 256, 258–259; Android, 147, 167–168, 170, 174–175, 182; driverless cars,

235–236; Google Earth, 209–210; Google Nest, 175
Gopher, 135–137, 138–139
Graphical User Interface (GUI), 104–105, 107–108, 115–116, 175
GRU (Russian Military Intelligence), 277–279, 281
Gunter, Edmund, 9, 10, 11

Hackers, 101, 145, 147, 247–252, 256–257, 259, 263–264, 266, 268, 271, 278–284
*Hackers* (movie), 266
Hagelin, Boris, 261
Hamilton, Margaret, 73
Hanke, John, 209
Harvard Mark I, 25, 38, 39–40, 56
Hastings, Reed, 155
Hawking, Stephen, 234
Hawkins, Jeff, 162–163
Heath Robinson, 36
Hellman, Martin, 252
Heron of Alexandria, 2
Herschel, John, 15
Hewlett, William, 99
Hewlett-Packard (HP), 77, 78, 94, 99–100, 104, 146, 159, 165
Hindu-Arabic number system, 5–6, 8
Hoff, Ted, 91–92, 93
Hofstadter, Douglas, 228
Hollerith, Herman, 22–25
Honeywell, 56–57, 65, 77, 80–81, 86, 96
Hopper, Grace, 40, 47, 64
Hutchins, Marcus, 283
Hyperlinks, 137
Hypertext, 106, 137–138
Hypertext Markup Language (HTML), 137–138, 141, 165, 212
Hypertext Transfer Protocol (HTTP), 137–138

Ibuka, Masaru, 149–150
Idestam, Fredrik, 161
Informatics, 85–86
Information Management Sciences Associates Incorporated (IMSAI), 97, 103–104
Information Processing Techniques Office (IPTO), 104–105, 121–123, 125
Institute of Electrical and Electronics Engineers (IEEE), 32, 49, 108, 222
Integrated Circuit (IC or microchip), 69–71, 72, 73, 75–76, 91–92, 150
Intel Corporation, 90–92, 93, 99, 101, 103, 110–111, 116, 146, 223
Interface Message processor (IMP), 124, 127–128
International Business Machines (IBM), 24–25, 39–40, 47, 53, 56–62, 63–67, 75–77, 79, 80–83, 84–87, 90–91, 101–104, 109–110, 114–117, 122, 125–126, 130–131, 132, 146, 163–164, 171–172, 226; internet, 130–131, 132–133; OS/2, 78, 115–116; personal computer, 101–104, 109–111, 114; SABRE, 66; Stretch project, 60–62, 90; System/360, 80–83, 85
Internet, 79, 105, 107, 114, 128, 129–131, 132–136, 138–141, 187, 190, 199–201, 205–207, 213–215, 216–217, 219–220, 241; backbone, 130–31; digital divide, 241–245

Internet Explorer (IE), 139–140
Internet of Things (IoT), 219, 222, 257
Internet Service Providers (ISPs), 130, 133, 187, 199, 272; AOL, 132–133, 135, 139–140, 187, 189–190; CompuServe, 132–133, 139, 187; Prodigy, 132–133, 187, 199
IP address, 128, 130–131, 195
Ive, Jonathan, 170

Jacquard, Joseph-Marie, 19–20
Jacquard Loom, 19–20, 23
Jelinek, Fred, 172
Jetons, 8
Jobs, Steve, 99, 104–105, 107–109, 165–166, 169, 247
Joy, Bill, 228

Kahn, A. Q., 269–270
Kahn, Bob, 128
Kao, Charles K., 220–221
Kardach, Jim, 223
Karim, Jawed, 201
Kasparov, Garry, 226
Kay, Allen, 105
Kelly, John, 171
Kelvin, Lord, 3
Kepler, Johannes, 10
Kernighan, Brian, 80
Keyboard, 33, 110, 158, 164, 167, 169
Kilburn, Tom, 44, 62
Kilby, Jack S., 69–71, 75
Kildall, Gary, 93, 103–104
Killer application, 101, 103, 108, 125, 154, 164
Kimsey, Jim, 132
Kircher, Athanasius, 11
Kittlaus, Dag, 173–174
Knots, 6–7
Kramer, Kane, 151
Kratzenstein, Christian, 171
Kremen, Gary, 206–207
Kroemer, Herbert, 71
Kurzweil, Ray, 227–228

Lamarr, Hedy, 259
Lebedev, S. A., 53–54
Leibniz, Gottfried Wilhelm, 13
Leibniz Wheel, 13, 14
Licklider, J.C.R., 79, 121–122, 124
Linux Operating System, 142, 144, 146–147, 167, 185, 237–238, 244
Lipstein, E., 15
Listserv, 130
Local Area Network (LAN), 133–134
Logarithms, 10–11, 15, 27
Lorenz Encoding Machine, 36–37
Lotus, 103–104, 112, 117
Lovelace, Lady Ada Augusta, 20
Lucas, Edouard, 11

Ma, Jack, 216
Machine (Binary) Code, 62–64, 88
Macintosh, 104–109, 110, 115, 117, 146, 152
Macron, Emmanuel, 281
Magnetic Core Memory, 53, 58, 60, 62, 91
Magnetic Ink Character Recognition (MICR), 59
Magnetic Tape, 47, 54, 58–59, 62, 85
Malware, 253–254, 257, 262, 264–267, 270–271, 283–285
Manchester Mark I, 43–44

Mannheim, Amedee, 11
Manning, Chelsea, 279
Markkula, Mike, 99
Markov, Andrey, 172
Massachusetts Institute of Technology (MIT), 28, 39, 52, 67, 73, 77, 79, 100, 108, 112, 121, 124, 139, 143, 171, 173, 229–230, 243
*Matrix, The* (movie), 228
Matsuzaki, Kiyoshi, 8
Mauchly, John W., 30–32, 40–43, 46–49
Mazor, Stanley, 92
McCarthy, John, 67, 79
Mechanical calculating devices, 11–14, 15, 17–18, 28, 41
Memory, 19, 29, 50, 57–58, 75, 83; disk drive, 60, 62; EPROM, 91; expansion card, 95, 97–98, 100; magnetic core, 53, 58, 60, 62, 91; magnetic tape, 47, 54, 58–59, 62, 85; mercury delay line, 45, 47, 53; optical drive, 108; punched cards, 18–19, 28, 30, 40, 45, 56–58, 61–62, 64–66; punched tape, 38; RAM, 19, 76, 91–92, 91; ROM, 91–92; rotating drum, 44, 58–59; virtual, 62, 78, 83
Menabrea, Luigi F., 20
Mercury Acoustic Delay Lines, 45, 47, 53
Metcalfe, Robert, 127–128
Micro Instrumentation and Telemetry Systems (MITS), 94–98
Microcomputer, 78, 93–104, 107, 109–110, 169
Microprocessor, 90–92, 97–98, 100, 102–104, 107–108, 110–111
Microsoft, 96–97, 102–104, 108, 109–110, 114–119, 134, 139–140, 146–147, 157–158, 167, 169, 174, 184–185, 191, 197, 209, 224, 238, 283–284; Azure Cloud, 185; Cortana, 175; DOS, 103–104, 110, 114–115; Encarta, 191, 197; Internet Explorer, 139–140; Kinect, 157–158; Office Suite, 117, 175, 224; software distribution, 137–138; Surface, 169; Windows, 78, 115–118, 134, 146, 165, 167, 169, 185; Xbox, 118, 157
Milgram, Stanley, 197
Minicomputers, 76–78
Minsky, Marvin, 66–67
Mitchell, John F., 160
Modem, 122, 129, 132, 181
Moore, Gordon E., 75–76, 90
Moore School of Engineering, 4, 41, 43–45
Moore's Law, 75–76, 110, 149, 219, 221
Morland, Samuel, 11, 13–14
Morris, Robert, Jr., 251
Morris Worm, 251
Morse, Samuel F. B., 122
Morse Code, 34
MOS 6502, 98, 100, 102
Mosaic, 138–139
Motherboard, 95, 231
Motorola, 97–98, 107–108, 159–161, 164–168, 173–174
Mouse, 106–107
MP3 music format, 151–152; players, 165

Mueller Investigation, 276, 279–280
Muller, Johann Helfrich, 15
Multiuser Dungeons (MUDs), 113–114, 187
Murdoch, Rupert, 199
Musk, Elon, 179, 234, 237–241
MySpace, 198–199
MySQL, 144, 203

Nadella, Satya, 185
Nakamoto, Satoshi, 262
Nanotechnology, 288
Napier, John, 10–11
Napier's bones, 10–11, 13–14
Napster, 152
National Aeronautics and Space Administration (NASA), 71–75, 106, 230, 238–239
National Cash Register (NCR), 24–25, 60, 65, 77
National Science Foundation (NSFNET) backbone, 130–131
National Security Agency (NSA), 61, 215, 248–249, 254, 257, 261, 272–274, 281–284
Negroponte, Nicholas, 243
Nelson, Ted, 137
NetBSD, 144
Netflix, 155–156, 184
Netscape Communications, 139–140
Network Control Protocol (NPC), 127–128
Network Working Group, 128; Request for Comments (RFC), 128–129, 131, 139, 223–224
Networks, 51, 71–72, 96, 105, 107, 113–114, 121–125, 126–131, 133–134, 137, 220–224, 225–226
Neural Networks, 225–226
Newmark, Craig, 212
Newton, Isaac, 13
NeXT Computer, 108–109, 137
Nintendo, 113, 158, 208, 210; Game Boy, 156–157; Pokemon/Pokemon Go, 207–208, 210; Wii, 158, 208
Nokia, 161–162, 164–165, 167, 223
Noorda, Raymond J., 133
Novell, 133–134
Noyce, Robert, 70–71, 90
NSF Computer Science Network (CSNET), 130
Numeration (counting), 5
Nygaard, Kristen, 89

Object-Oriented Programming (OOP), 89, 105, 108, 137
Odhner, Willgodt T., 14
Olsen, Ken, 77
Omidyar, Pierre, 188–189
Open bus architecture, 86, 89, 94
Open Source, 142–147, 185, 192–193, 224, 237–238, 244
Operating system, 62, 78–82, 88, 103–105, 109, 110, 114–117, 143–147; DOS, 103, 104, 107, 109, 114–117, 173; Linux, 142, 144, 146–147, 167, 185, 237–238, 244; OS/2, 78, 115–116; SCP-DOS, 103; UNIX, 78, 79–80, 108–109, 116, 134, 139, 142–144, 146; virtual memory, 62, 78, 83; Windows, 78, 115–118, 134, 146, 165, 167, 169, 185
Optical drive, 108

Optical fiber cables, 184, 220–222
Organum Mathematicum, 11
Orrery, 2–3
OS/2, 78, 115–116
Osborne Computer, 104
Oughtred, William, 11

Packet switching, 123, 129–130
Page, Larry, 142
Pager, 163–164
Palm, 162–165
PalmPilot, 162–163, 225
Papert, Seymour, 243
Parker, Sean, 152
Pascal, 89
Pascal, Blaise, 12–13
Pascaline, 12–13
Patriot Act, 272–274
PayPal, 140, 190, 237–238
Perot, H. Ross, 87
Personal Computer (PC), 78, 92, 96, 101–104; clones, 83, 109–111, 112, 114–115, 117
Personal Digital Assistant (PDA), 158–159, 162–169
Philco, 56–57, 61
Philon of Byzantium, 2
Phishing, 254–256, 271
Pitroda, Satyan, 159
Pixar Animation Studios, 109
Planimeter, 3
Playfair, William, 210–211
Podesta, John, 279
Pong, 113
*Popular Electronics*, 93, 95, 96
Porter, Robert P., 23
Postel, Jon, 129, 206
Postley, John, 85
Prime Computer, 77
Printer, 59
Prodigy, 132–133, 187, 199
Programming, 19–20, 39–40, 45–46, 60, 62–64, 67, 80; BASIC, 87–89, 96–97, 101, 103–104, 114, 158; C, 80, 89; C++, 89; COBOL, 64, 87; FORTRAN, 63–64, 72, 88; Java, 89; JavaScript, 138–139; Machine (Binary) Code, 62–64, 88; Report Program Generator (RPG), 64; structured, 87–89, 105
Project Whirlwind, 52–54
Punched cards, 18–19, 28, 30, 40, 45, 56–58, 61–62, 64–66
Punched tape, 38
Putin, Vladimir, 276

Quadrant, 9
Quipu, 6

Rabdologia, 10–11
Radio Shack, 97–98, 159
Railway Clearing House, 22
RAND Corporation, 65, 123
Randolph, Marc, 155
Random Access Memory (RAM), 19, 76, 91–92
Ransomware, 263–266; NotPetya, 284–285; WannaCry, 283–284
Read-Only Memory (ROM), 91–92
Read-Only Memory Basic Input/Output System (ROM BIOS), 110
Red Hat Linux, 144
Rejewski, Marian, 34
Remington Rand, 24–25, 47–48, 56–57, 89

Report Program Generator (RPG), 64
Reynolds, Joyce, 129
Rifkin, Stanley Mark, 250
Ring, Douglas, 160
Ritchie, Dennis M., 79–80
Rittenhouse, David, 2–3
Rivest, Ron, 252
Roberts, Edward, 93–98
Roberts, Larry, 124–126
Robotics, 229–234
Rotating Drum Memory, 44, 58–59
Royal Society, 13, 16, 17
Royal Society Computing Laboratory, 37, 44
RSA (Rivest, Shamir, Adleman) Encryption Algorithm, 252
Russell, Steve, 112
Russell, Stuart J., 234
Rydbeck, Nils, 164–165

SABRE project, 66
Sanders, Bernie, 278–279
Sanger, Larry, 192–193
Satellites, 121, 128, 155–156, 220, 238–240, 243, 251, 258, 260
Scheutz, Georg and Edvard, 17–18
Schickard, Wilhelm, 12
Schmidt, Eric, 183
Schott, Gaspard, 11
Schreyer, Helmut, 38
Science fiction, 94, 142–143, 146–147, 149, 227–228, 232
Search engines, 141–142, 183, 217, 229
Seattle Computer Products DOS (SCP-DOS), 103
Section 230, 199–201
Sector, 9–10
Sega, 113
Semi-Automatic Ground Environment (SAGE), 52–54, 56, 61, 65, 122
Semiconductor material, 55, 70, 76
Shadow Brokers, 282–283
Shamir, Adi, 252
Shockley, William B., 55–56, 70
Silicon Graphics (SGI), 90, 139
Simonyi, Charles, 117
Skoll, Jeff, 188–189
Skype, 190
Slide rules, 11, 27, 28
Smith, Adam, 21
Smith, George, 221
*Sneakers* (movie), 266
Snowden, Edward, 254, 272–274, 282
Social media, 197–199
Software, 65–66, 84–87; bugs, 145; distribution, 118; open source, 142–147, 185, 192–193, 224, 237–238, 244
Solomon, Les, 93–94
Sony, 113, 150, 181, 259; Betamax, 154; Discman, 151; Walkman, 150–151
*Spacewar!*, 77, 112–113
SpaceX, 179, 237, 239–240, 243
Sperry Rand Corporation, 32, 48, 59, 62, 77, 84, 89–90
Spreadsheet, 100–101, 103–104
Stallman, Richard, 143–144
Stanford Research Institute (SRI), 59–60, 106–107, 124–125, 137, 173; ARPAnet, 107, 124–125, 129–130
Stoll, Cliff, 247–249, 252
Stored Program Technique, 45

Strachey, Christopher, 79
Stratonovich, Ruslan L., 172
Stretch Project, 60–62, 90
Structured Programming, 87–89, 105
Stuxnet, 269–271
Suess, Randy, 131–132
Supercomputers, 62, 89–90
Surveillance capitalism, 212–213
*Swordfish* (movie), 266
System/360, 80–83, 85

Tables (calculation tables and tablets), 3, 9–10, 21
Tablet computers, 169–170
Tabulating Machine Company (TMC), 23–24
Tally sticks, 7; Exchequer tally system, 7; Ishango bone, 7
Taylor, Frederick W., 21
Taylor, Robert, 104–105, 123–125
Telephones, 122
Tesla, 236–237, 239–241
Texas Instruments (TI), 69–71, 94, 149–150
Thi, Andre Truong Trong, 96
Thompson, Ken, 79–80
Thomson, James, 3
3Com, 127
Tide predictors, 3
Time-Division Multiple Access (TDMA), 127
Time-Sharing, 62, 78–80, 87–88, 101, 122, 132, 183
Tomlinson, Ray, 125
Tomlinson, Roger, 209
Torvalds, Linus, 144
Toshiba, 110
Transistors, 54–56, 61, 69–71, 76, 149
Transport Control Protocol (TCP), 127–129
Transport Control Protocol Internet Protocol (TCP/IP), 127–129, 130, 134, 139, 220
Trump campaign, 276, 279–280
Tsai, Joe, 216
Turing, Alan, 35–37, 44, 57, 66, 227
Turing Machine, 35–37
Turing Test (AI test), 66, 227
*2001: A Space Odyssey* (movie), 171, 174, 228
2016 Election, 205, 258–259, 274–281

U-boats, 32
Unisys, 48
UNIVersal Automatic Computer (UNIVAC), 46–49, 56, 57, 58, 59, 61, 62, 89
Universal Resource Location (URL), 137–138
University of California, Los Angeles (UCLA) and ARPAnet, 124–125, 128
University of California, Santa Barbara and ARPAnet, 124
University of Utah, 112–113; and ARPAnet, 124
UNIX, 78, 79–80, 108–109, 116, 134, 139, 142–144, 146
Usenet, 134–135, 152, 187

Vacuum tubes, 29, 37–39, 41, 54, 57, 61, 149
Vault 7, 281–285

Vernam, Gilbert, 36
Videocassette Recorder (VCR), 153–155, 207
Video Home System (VHS), 154–155
Vinge, Vernor, 227
Virtual Memory, 62, 78, 83
Virus, 251, 266
VisiCalc, 100–101, 103–104, 108, 109, 112
Vlacq, Adrian, 10
Voice User Interface (VUI), 170–176
von Neumann, John, 42–43, 45–47, 57, 66
von Neumann, Klara, 46
von Neumann architecture, 43

Wales, Jimmy, 191–193
*WarGames* (movie), 97, 266
Watson, Thomas J., 24, 39
Watt, James, 11
Weinreich, Andrew, 197
Wensley, Roy J., 170–171
What You See Is What You Get (WYSIWYG), 105
Whitworth, Joseph, 16
Wiberg, Martin, 17
Wiki, 192
Wikileaks, 279–280, 281–282
Wikipedia, 191–197, 229
Wilkes, Maurice V., 45
Williams, F. C., 44
Williams Tubes, 44, 57
Wingate, Edmund, 10
Wireless Networking, 126–127, 222–223
Wirth, Niklaus, 89
Wood, Thomas, 8
WordPerfect, 117, 224
Word Processing Program, 101, 104, 117, 224
World Wide Web (WWW), 109, 137–141, 229
Worm, 96, 251, 266, 270
Wozniak, Steve, 98–100, 234, 247
Wu, C. F. Jeff, 211

Xerox Alto, 105
Xerox PARC (Palo Alto Research Center), 104–107, 115, 117, 127–128, 133–134, 180–181, 209

Yahoo!, 141, 188, 203, 267
Yang, Jerry, 141
Yong, W. Rae, 160
YouTube, 201–202, 205, 216, 276

Zennstrom, Niklas, 190
Zilog, 97–98, 104
Zuckerberg, Mark, 202–203
Zuse, Konrad, 37–39
Zuse Computers, 37–39

**About the Authors**

**Eric G. Swedin** is a professor of history at Weber State University in Ogden, Utah. His publications include numerous articles, seven history books, four science fiction novels, and a historical mystery novel. His book *When Angels Wept: A What-If History of the Cuban Missile Crisis* won the 2010 Sidewise Award for Best Long-Form Alternate History.

**David L. Ferro** is a professor of computer science and the dean of the College of Engineering, Applied Science, and Technology at Weber State University in Ogden, Utah.